Horn Stew

BY

Darol Dickinson

MMXVIII

Copyright © 2018 by Fillet of Horn Publishing
All rights reserved.

Darol Dickinson, Author
Darol Dickinson, Photographer

Printed and published by
Fillet of Horn Publishing
Barnesville, Ohio U.S.A.
First Edition, First Printing

Printed in the United States of America

ISBN: 978-0-9753077-4-8
Library of Congress Control Number: 2018903256

Dedication

To my dad, Frank Dickinson,
1917 – 2005

who said:

"Never invest in anything you can't throw a rock at."

"If you ever have money, invest in land and cattle. The government can't print any more land, and your cow will have a calf in the spring, then you have two."

"Earn your own money for college — then you will remember what you learned."

"I never hired a cowboy that I didn't learn how to do something better, or more often, how to never do it again."

"Listen to everyone, then do as you darn please."

"If you don't go to their funerals, they won't attend yours."

"Son, always put money in the church collection plate. I always put in a dollar, no matter if it's a good sermon or not."

Frank Dickinson

Contents

Introduction .ix
Gorgeous George the Steer 1
The Burials of Silky Fox 5
How Not to Attend the AQHA Convention. 18
Vessels and Go Man Go 20
Ruidoso Downs, New Mexico 22
The Johnny Cash Longhorn Herd 26
Appaloosas and Carl Miles 33
Bogus Bonus 38
The Rest of the Butler Story. 41
Guthrie Buck and Walt Hellyer. 56
Wallaby 58
Deep Bob and Empty Pockets 61
Don Quixote's Mystery 63
Tonto Bars Hank and Walter Spencer 71
Hard-Water Christmas 75
Dealing with J.L. Collier. 77
The Bob Shultz Bull 81
Joe Queen, Rogers, and Murphy 85
The Shadow Saga 88
Moolah Bux. .103
Rhonda K and Silkay—The Process 105
Hauling With or Without Class 110
The Partlow Purchases. 115
He's a Dude of Houston 130
Working Big Pastures 134
Three Bars, Vail, and Merrick. 137
"King" 142

Splash Bar and Michael Mulberger 146
Starting the Nevada Herd 150
Frank Doherty and Doherty 698 153
The Pretty Penny Ranch 158
Hank Wiescamp—The Legend 164
The Tahitian Cattle Trade 171
Impressive. 176
BRY Squeeze Chute .181
Titan Wolf of Kooskia 186
Drag Iron 190
Clear Point—The Linebred Outcross. 193
Power Game—The BueLingo 196
Field of Pearls, the Family. 199
Dickinson Cattle Company—One Month 204
A Word of Praise 215

Introduction

I started the business of taking photos and doing portraits of livestock before I had a driver's license. The first time I went to Hank Wiescamp's to do the first Skipper W painting, my mother drove me over the mountains to Alamosa. Once, she drove me to Sarcoxie, MO, to do sketches and photos for a portrait of the Appaloosa horse Coke Roberds and Jim Wilde's National Champion, Star Mist. Early on, I took a train from Colorado Springs to south Texas to do photos. I borrowed a car from relatives or friends once I got a license, to do work in the San Antonio and Dallas areas. As things progressed, I recall my first plane ride from San Antonio to Colorado Springs on a propeller plane, before jets. Most planes were not jets yet. The business of photographing livestock required travel to the animals.

My high school principal, Rance English, introduced me to his partner, a photographer—Bob Hagen. I was a senior and wanted to learn about photography. Bob walked me through simple basics. We went to some sports events and practiced action-photo skills. One day, we decided to photograph a rodeo. I had a Minolta 35mm with a standard lens. To get a good sharp shot, I had to be in the arena, and sometimes between the bull and the clown. As the rodeo progressed, I shot roll after roll. My running around the arena and occasionally up a fence was an exasperating day. When the last bullride was over, we went to Bob's photo darkroom and started developing film. I was tired.

All photos were black-and-white. An 8"x10" sheet of Kodak photo paper cost 3¢. About midnight, all the film was developed and we started printing photos. I was tired. Bob said we had to get this done. We worked in the dark room for hours. I was tired. At first light, I needed to get a little rest. Bob said to just take a cold shower and we would get back to the rodeo grounds and sell photos. Bob's wife, Joann, fixed breakfast, then off to the rodeo. We displayed the photos on the hood of Bob's old car surrounded by cowboys. We sold photo enlargements for $1 each. They cleaned us out, except for a few that were out of focus. When the rodeo started, I was back in the arena taking photos. I was tired. As Bob sold photos, he got every cowboy's address. Photos taken the second day were mailed to eager contestants with an invoice, which most of them paid. In summary, we grossed something over $400 in 2 days, divided it up, and $100 for a day's work for each of us was like robbing a bank in 1959. We learned that the better the photos, the more reprints were sold. However, an equally great lesson learned was that I could work a 40-hour day and it was very profitable. Later in life, I realize the knowledge learned from Bob Hagen was that long days were amazingly profitable. This was one of the greatest lessons of my entire life. Had I weakened for an 8-hour day, that decision I believe would have cut my lifetime achievements in half.

My first camera was a Brownie Hawkeye, then a Minolta 35mm, then up to the big league with a Speed Graphic 4"x5" box camera. I had one Speed Graphic with a German Ektar lens which served me well to do front views. I used a 5.25" lens for side and rear shots. Ronda, my mother, made a sturdy yellow vest with pockets of all sizes for photo equipment. I could switch from cut-film to a roll-film adapter then to the Ektar. One roll-film adapter would have color and one black-and-white. Most ads in the '60s and '70s were not color, so most photos were black-and-white. Color was very expensive and only for magazine covers and postcards. Later, color got more economical. I used two Speed Graphics and one Super Speed Graphic. I carried all these big cameras everywhere, dozens of rolls of film, and also a little

Calendar or magazine covers were taken mostly vertical. Linda was the main model for covers as horses and Longhorns were wrapped around colorful scenery.

35mm camera with a long telephoto for head shots and rodeo stuff. It required two large hand cases to get it all transported. The wonderful zoom lens was not perfected to professional sharpness back then.

With the multiple sizes of roll film, it was slow changing. Some rolls had 8 shots, the sheet film had 1 shot, and at most a roll of 35mm would be 37 shots. I was changing rolls continually. Often the livestock would move, then it was a start-all-over effort.

When the right setting was located, I set up magazine cover

shots. Years ago, magazines would pay a nice fee for a cover photo. Covers were the classy introduction to magazine content. I would drive the country hunting a scenic setting, then arrange to move a herd of livestock over or through the setting for cover shots. This was done with 4"x5" ektachrome sheet film. Today, a magazine staff person just takes a photo out the car window of a cow in the pasture with a fence post, and that becomes an economical cover. Some publications auction or sell the cover. Most of the art qualities of livestock-magazine covers are gone.

These big cameras were fragile. If something got bumped or banged around on a trip, I was out of business. Therefore, I refused to baggage-check any of the camera boxes. I carried them all personally.

This photo is the wonderful old slow Speed Graphic with the Ektar lens. If you did not capture the shot at the exact time, it was slow to crank it back for the next shot; it could be too late. I used to buy most of my cameras at hock shops. I stopped by hock shops all over to pick up cameras that were half the price of a new one. Today I have a Canon EOS Rebel T2i with a zoom lens 28–135

mm. It does everything the two boxes of cameras did, and film is replaced with a tiny computer card that holds over 800 shots. Today, there is no excuse for not taking great photos. It is a piece of cake. It is a walk in the park with auto-focus, virtually no cost per shot—just blast away hundreds of shots and delete the bad ones. Anyone can do it.

In 1979, I wrote the book *Photographing Livestock*. It identified the 7 errors in photography and taught livestock owners to take their own photos. After the book was published, it served as a textbook for the *Dickinson Livestock Photo School*. This school was produced 28 times for 22–30 students at a time between 1979 and 1985. Gary Lake, a dear friend and talented photographer, worked with me for many of the schools. The book published by Northland Press sold out 5 printings. Although it is old, the process of taking correct photos is still correct, yet the info about equipment is foreign to modern photographers.

This simple profession of photography and oil portraits of livestock had a wonderful reward that many would kill for—I got to see most of the great livestock in the nation. I was welcomed to ranch homes for dinner, traveled the back trails of famous ranches, and some people shared their personal problems with me. I got to know a lot of people on a close-friendship basis. I had run of the King Ranch, Vessels, Merricks, Wiescamps, and the 6666 Ranch. I attended the first 16 *American Quarter Horse Congress* events, seeing all the great horses. Hopefully you will enjoy reliving these travels, the history of unique people and places—plus the current Texas Longhorn industry—even more wild and crazy unique people.

Things change in 60 years. Anyone can take great photos now, if they know their subject and just go do it. Most people today are not persistent.

After publication of the book *Fillet of Horn*, there were several dozen chapters that fell through the cracks—some too long, some too short. A book can just be so big, then you have to carry it in a wheelbarrow. As *Fillet* is just about out of print, it is time to pick up the other historic pieces and cook them into **Horn Stew**.

All chapters are new. There are detailed sightings when photographing famous World Champion horses with personal notes—like Three Bars, Joe Queen, Silky Fox, Impressive, Silkay, Moola Bux, Prince Plaudit, Go Man Go, and Tonto Bars Hank. Personalities attached to these legendary livestock include Hank Wiescamp, Walter Merrick, Fennel Brown, Tommy Manion, J.L. Collier, Sam Partlow, Michael Mulberger, Carl Miles, Audie Murphy, Blackie Graves, Walter Spencer, etc.

Mixed in with the horse industry are historic accounts of famous Texas Longhorns like Bogus Bonus, The Shadow, "King," Classic, Don Quixote, Doherty 698, Tempter, Drag Iron, and Clear Point. Lessons learned include promotional ideas that worked or didn't work, people that you would love to meet today and continue their friendships, and also some people you would be scared to shake their hand.

Some chapters are just a single page and some are 20 pages. The record will document crazy stuff, interesting stuff, disgusting stuff, hard-nosed negotiations over peanuts, really quality nice people, and events that I wish had never happened. You will laugh at me for being so stupid, but then, it may help you not to be as ignorant yourself.

The studio dark room was a place for night work. The door was locked so no one would come in and ruin the photos; the phone quit ringing. Lab work took hours, but it was okay.

The chapters are "sort of" chronological. Some start back in 1955, then a lifetime of connective tissue drags you kicking and screaming up to the current date.

Travel from Canada to California to Florida is normal. Our business has been so bad it took the whole world to make it happen. My first main income was from photographing livestock, then doing oil portraits of great livestock and book illustrations. The last years have been diversified around ranching, raising of registered cattle, and selling and exporting of embryos, cattle, and bull semen world-wide. The chapters bounce in and out of these and related efforts.

Some chapters may not be entertaining but do contain methods of evaluating scientific genetics. Ways to mate bloodlines to progressively move livestock breeds forward is a very profitable study. Raising better genetics, working to stay on the front cutting edge, these are challenges of a most strenuous nature.

Dad was always ready to bring Silky Fox to pose for the Dickinson/Lake School of Livestock Photography. Every morning, 6–12 horses or cattle were brought in. The owners got free photos. Students included magazine editors who wanted to learn livestock photography. We had two Miss Rodeo America students.

If coachable, the education from talking to highly motivated, successful achievers is invaluable.

In *Horn Stew,* you will see vast differences in people. There is a special sweat-equity art attached to the results of producers who love their tasks—who go beyond a profit/business and endure the hard and painstaking challenges of being a superior achiever. If one is raising great chickens, hogs, horses, cattle, or even taking photos, observers can see the difference when one is committed by a love of the art rather than just trafficking in a commercial product. You will see the difference between a breeder and a trader. One may have very long-term goals that help a lot of people, and others just do things for an immediate quick buck.

My tortured wife, Linda, has changed diapers in many states, in hot or cold weather. Our four children appear surprisingly normal to us. All are in some way involved in the ranch business today. You will find chapters when they were young, and also in current adult times. My oldest son, Kirk, did more work laying out and preparing this book for print than the time I spent writing it. Without his dedicated skills, I would never have attempted this tribulation.

If you are old or new to livestock breeding, ranching, and the livestock business, you will enjoy the flavor of *Horn Stew*. There will be healthy vegetables cooked to perfection with the strong protein of real horn all mixed together, perhaps with some juicy morsels of high Omega-3 steak. If you like ranching, people, and rural things, *Horn Stew* will taste real good. Pull up a comfortable chair. Grab a large napkin—you will need it. This may get sloppy!

> *"Far and away the best prize that life offers is the chance to work hard at work worth doing."*
>
> ~ **Theodore Roosevelt**

Gorgeous George the Steer

The Barnesville Livestock Auction is held every Saturday. My best old friend Bill Farson attends every sale if his health allows. I think he is 86. He calls me from the sale barn if there are some good Texas Longhorn hamburger cattle selling. Sometimes he picks up 1 or 2 which go into the ranch store grind inventory. Bill called Saturday, September 13, 2014, and said there was a Longhorn steer going through the auction. Did I want Bill to buy him? He said the steer was young, not over 5 years and had horn about 7' wide. He thought he would bring about $1 per lb. I said, "Buy him!"

Bill is one of the few experienced cattlemen who can guess a critter's weight within a few pounds. He has the "eye." He has watched millions of cattle come into the sale ring, watched them weighed, and knows what he is seeing—a real cowman. Bill can see a critter for 30 seconds and describe him to you for 10 minutes. That is the eye of a authentic cattleman. It comes from a lifetime of buying and selling cattle. You don't read this stuff out of a book or learn it in college. You learn it by spending years in dusty old spit-bucket cattle-auction barns.

Bill called about 4:00 PM and said, "You own a steer." I asked, how much? Bill said $1.59. That made him cost $2,176.20. He was the high-selling critter of the day. In fact, when he came into the ring the auctioneer stopped the sale, stood up, and took a photo of him. Others came down to ringside and took photos. That never happens at Barnesville. Never had happened before.

I was anxious to see this big steer. Joel took a ranch check to pay out, brought him home, and put him in a pen with some good Ohio hay. The steer was quiet and loaded easy. You could walk around him and he didn't worry about anything. We stood and looked at this beautiful steer, trying to figure out his origin. Every family of cattle has unique points of anatomy. With careful evaluation, certain body parts reveal their identity. I told Joel he looked like a son of our old bull Superb. Joel thought he had some Zhivago blood because he had above-aver-

Bill Farson has watched millions of cattle sell. He knows their weight, current value, and where their next owner lives.

age bone. I could see a trace of what I thought was Superb in the shape of the nose bones. He was measured with an impressive 85¼" tip-to-tip. It was a long shot that a young steer in eastern Ohio would have this excellent horn, size, bone, and substance from any genetics other than Dickinson Cattle Company, LLC (DCC). He had no brands. With careful observation, we saw that he had two small holes, one in each ear, where DCC normally places an ID tag and an insect-vapor tag. We were certain he was a steer DCC had bred and sold for freezer beef as a weanling.

Linda, my wife, does all the DCC registrations and knows our records like a rabbit knows the briar patch. She was determined to find this steer's origin and checked every Superb steer calf. He looked too young to be a Superb steer, because we had sold Superb too soon to be his sire. Every calf born at DCC is photographed and recorded on the day of birth and entered into the computer. Linda went through every photo until she found the perfect spot-pattern match. It took her about 4 hours, but she was viciously persistent.

The 85¼" steer was born on 5-3-2010. He was sired by the DCC bull Fixer and out of Lynx Mountain, who was a daughter of Superb. In checking his pedigree, we found that he was linebred Zhivago tracing back 5 times to the big bull. He was sold with 4 other little steers at early weaning for $360 each to a fellow named Douglas, who lived about 3 hours away down in southern Ohio.

I called Douglas and asked him about the steer. Douglas said he had sold him to a neighbor lady who kept him as a pet. He was halter-trained and had been ridden a little. I told Douglas I had bought him. Douglas was shocked, "No!" He said, "You couldn't have!" Douglas said the neighbor lady had run out of pasture and sold him very cheap to another neighbor who promised to keep him forever as a pet for his kids. He would have a loving home—and that clinched the deal. In fact, Douglas, had met the buyer and helped him load the steer with a bucket of feed about 8:00 that very morning. (Barnesville is about a 3-hour haul away.) The lady did not get much for the steer but mentioned that he ate a lot of feed.

A few hours after the steer sold, photos began to appear on Facebook. It was reported that when he entered the sale ring, he turned his head at an angle to get inside. The seller raised his hand to testify about his steer. He recounted, in almost total tears, how he had found the steer as an orphan near death. His family loved this kind and gentle steer, his kids had nursed him to health and raised him on a bottle, he was the family pet, it pained him considerably to let him go, but for very personal reasons he had to sell him today. Seller's remorse was reportedly

Gorgeous George is a big, beautiful, and gentle steer. At maturity he will go well into the 100" tip-to-tip range.

Gorgeous George the Steer

expressed all during the bidding. No doubt it hurt him bad to part with this dear family friend. The more remorse he described, the more Bill had to bid. It was emotional. Some reported there were sympathetic tears in the audience.

We identified all the history and got the steer's registration in order. He is now correctly registered and given the registered name "Gorgeous George." Bill misguessed him as he wasn't 7'; he was 85¼" tip-to-tip (T2T), but he was under age 5. Normally Bill doesn't miss that far. And who would guess the passionate orphan story would jack the price by $59 per hundredweight.

Gorgeous George fattened up fast, enjoyed a short stay at DCC, and was priced for $9,000 on the ranch sale internet inventory. He was a nicer steer than any on the ranch, and he was quickly purchased by a pasture-art collector in Texas. Now Gorgeous really does have a loving home, with plenty of grass and honest credentials. And, also, he was never really a starving orphan!

"If you hang around bank robbers, you will soon be driving a get-away car."

The Burials of Silky Fox

Frank Dickinson, my dad, was born June 21, 1917. During his years in school, he sat by a large window and watched a nearby rodeo arena. It was a place where locals roped calves, rode horses, and dreamed of being real western cowboys. There was no John Wayne or Clint Eastwood back then. It was even pre-Hopalong Cassidy times. The most famous cowboy in those days was Will Rogers.

Dad watched and tried to learn from others who watched one another try to learn. His first horses were draft horses used to plow, pull buggies or wagons, and do the things that were later taken over by John Deere. When Dad came to visit in Ohio over 70 years later, he loved to go to Amish country to watch the buggy horses. One time he saw 20 or so horses harnessed to buggies and tied to a parking-lot hitch rack near a large grocery store at Sugar Creek. "Hey, let's go check those horses out," he said. We parked near the row of tired, sweaty horses that Amish women had driven to town. The Amish women rise at pre-dawn, get their husbands off to work and kids off to school, then harness up and trot to town. The local hitching racks are the conversation places where you learn all about what is going on.

Dad wanted to know about the horses. He wanted to check every harness buckle, see how the hames were attached, and ask each woman where her horse came from, the breed, how old, etc. His experienced eye would catch any unique variation of anatomy or even how the hoof clip was shaped. His visits with the Amish were enjoyable, and they in turn appreciated his interest and love of fine horseflesh.

Dad was never blessed with abundant funds. Once during some hard times, he said his goal was to save $1 per day. He was raised through the Great Depression and never wanted that kind of torture again, nor for any of his family. The key was hard work and a cunning evaluation of anything that could be improved on or

While Lynn Anderson's song "Rose Garden" was topping the country-music charts, her heart was focused on Silk Worm. Winning the *All American Quarter Horse Congress* was the toughest competition in the Quarter Horse world. Lynn and Silk Worm were a brilliantly coordinated machine. Harold Campton photo

made better. He traveled the country, went to every major horse show, and searched the nation for quality bloodlines.

Dad was tender-hearted. He blamed it on the genetics from his father. He said his father would cry watching a stranger catch a train. Once Dad had a favorite dog hit by a car and killed. He dug a grave, buried him, and cried over the dog for hours. Every day he would stand by the grave for a while and shed some tears. After a week, Mom told him to get a grip. She said, "You wouldn't cry that much if I died!" I don't recall if that gripped him up or not.

"Conservative" was a word that described most things about Dad, but "tight" was more descriptive. He would drive all around town hunting gas at the lowest price; but when it came to buying competitive Quarter Horses, the whole deal was about quality. Price was not the only deciding factor.

In 1965 my sister, Vicky, and her husband lived at Big Bull Ranch, a large Hereford and Quarter Horse operation located on the south slope of the Canadian River, in the north Texas panhandle. As a result of a midnight-predator/barbed-wire encounter, a number of horses were cut up and even killed. One silky-haired chestnut colt was in a near "give-up" condition from a deep gash in the right shoulder rendering him unable to manipulate his right front leg. Vicky took on the job of daily care, medication, and the ordeal of doctoring. As the foal improved, a bonding developed with Vicky. Then over time, he healed enough to walk and went back with the pasture herd.

Somewhat over a year later, Dad purchased this same chestnut colt for $2,400 in the annual Reed Hill Ranch sale. His dam was a race futurity winner and Grand Champion show mare named Ima Pixie by Custus Rastus TB. His sire was Rapid Bar TAAA, a son of leading sire Three Bars TB. The chestnut foal was

Silky Fox was the only one of 34 paintings for *Color of Horses* in which markings were allowed. I argued with Dr. Green until he allowed me this one indulgence.

named Silky Fox. The auctioneer and the breeder both thought this yearling a long-shot considering the marked handicap of his huge injury. But not Dad. With the colt's great pedigree and a special spark that only Dad sensed, the great stuff was all right there.

Silky Fox grew and developed to 15-2 hands. He had the Thoroughbred frame with the exact correct Quarter Horse muscle. Dad enjoyed watching him develop year after year. He trained him to work cattle and do the basics of ranch work. As he matured, Dad bred him to his select mares, and the foals begin to arrive.

When Silky was 4, he was sent to roping school with Allan Johnson, one of Dad's friends. He learned fast and didn't have a mean bone in his body. He was shown at halter and won numerous shows, including Grand Champion at the Colorado State Fair. After breeding season, Dad took Silky to Billy Allen, a respected showman, for serious competition. Within 35 days, Silky Fox earned an AQHA championship with points in 5 categories, including halter, all at the same time. Dad was pleased but didn't seem shocked. That was the way he thought it would be.

As years went by, Dad stood Silky to the public for breeding service. The fee was $1,000, and mares came from around the nation to have a foal by Silky. Most progeny were bright chestnuts, often with white-stocking-feet décor. Silky's sons and daughters were winning in every halter, performance class, and even in the rodeo arenas.

Silky Fox, rear view

Dad and Silky were an item. Everywhere Dad went he had Silky Fox postcards in his shirt pocket and freely distributed them to all. He would book about 40 mares each year, and Silky made the ranch payments. Dad had a system down where he handled the mares and Silky alone, doing the breeding thing and even artificial insemination of multiple mares all by himself. Just him and Silky making a living selling the grand genetics. Dad was tight and seldom ever had an employee to help. Of course my Mom, Ronda, got drafted into every ranch job that required two people. She kept the records, filled out the registration applications and show entries, and held it all together.

Each time Silky or his progeny won an award, Dad carried the trophy. Behind the scenes was Mom, who more likely carried a shovel. Some days, several horse trailers would come into the driveway where mares were unloaded for Silky's court. As horses ride the Interstates, they freely relieve themselves at will, hourly enlarging a stack of horse-apples on the rear trailer floor. Mare owners always stopped right between Mom's back door and the steel horse barn. As the mares backed out, the trip's entire accumulation slithered randomly into the driveway about 20' from the house. Mom got tired of it—day after day. In minutes, she would shovel the apples up and relocate them into the flower beds. She was always known for her rich garden, lawn, and beautiful flowers, but shoveling was not something she enjoyed doing. Dad would show the clients Silky Fox and his foals and do a 2- or 3-hour ranch tour. When he finally came to the house, his lunch would be cold, but he often brought a load of hungry horse people with him. He expected Mom to read his mind, which didn't always happen, and to have a feast ready, with a smile—which also did not always happen.

The Quarter Horse business is known for being a rich person's plaything. Although Dad didn't drive around with a half-million-dollar shiny truck and trailer that the whole family could live in, he put his investment into great mares, and of course Silky. He wanted to be competitive in the arena and not necessarily in the parking lot.

As Silky's progeny started to blossom, other people were evaluating his potential for their breeding use. Matt Browning of Browning Arms Company, son of inventor John M. Browning, came during a Colorado blizzard to see the stallion

The achievement of making a plan come together, a plan that started prior to conception and carried out for dozens of years, climaxed when Silkay was All-Age Grand Champion Mare at the *National Western Stock Show* in Denver. It is one thing as a breeder and producer to buy a champion and a much harder thing to engineer every point of anatomy—two achievements worlds apart.

and considered buying him. Dad priced Silky beyond what he wanted to pay. (The Browning automatic arms were mounted on every U.S. plane and tank during World War II. The Al Capone mob's weapon of choice was the M1918 Browning Automatic.)

The famous Four Sixes (6666) Ranch owner, Ms. Anne Valliant Burnett Tandy, wanted to buy Silky. She left King County, TX, in the ranch plane, stopped by Sayre, OK, and picked up the noted horseman Walter Merrick. Walter was a friend of the family and consulted for "Miss Anne." He recommended that she purchase Silky, but again, Dad was above their price range. Later, Silky Fox would produce more AQHA point-earning get than any sire ever raised by the 6666 Ranch.

Dad was a lover of great horseflesh, and Silky was all his. If things didn't go well for Dad, you could find him seated on the hay manger in Silky's pen, just looking at him. Sometimes the dog joined them. It was a place of peace, of satisfaction, and quiet hope.

Mom once said that Dad was so close to Silky that when the day came that Silky died, Dad would just automatically die the same day. He laughed about it, never thought it would happen, but also never denied it. Dad trimmed Silky's feet, brushed the dirt off him, and measured out his rations exactly to a specified amount every day.

Dad had heard some horrible stories of horses burning up in barn fires, so we built a large barn with solid steel stalls. There wasn't a stick of wood in the whole building. He and I welded on it all winter to be complete by breeding season. It may be the only solid-steel horse barn ever. The walls were heavyweight, smooth sheet steel. It never needed a single repair because nothing ever broke. It had 12 stalls and exercise runs out the side with huge elm shade trees for Silky on a hot day. He had the front stall but was visible only from the inside of the barn or when he was out in his exercise paddock.

While us kids were trying to grow up, Dad milked dairy cows morning and night and then did the ranch business with cattle and horses. Mom had a job as seamstress at the largest wedding-gown store in Colorado Springs—Neufelds. Her job was to make a bride look her best for the big day, so when an odd-shaped one showed up, Mom would adjust everything to make her look as good as possible, considering the problems at hand.

One day Neufelds remodeled and removed a huge front door with 1"-thick shatterproof glass about 4' wide and 9' tall. When the workers stacked it in the trash for pick-up, Mom asked if she could have it. Dad and I raced to town before trash collection and picked up the door. We hauled it home, cut a hole in Silky's suite, and installed the door horizontally. Now Dad could see him all the time from the kitchen window through a recycled glass door.

Dad would get up at 4:30 AM and go to work. After he sold his milk cows and horses, he still got up at 4:30 AM and waited on the sun. If cattle work was needed, he went until it was done. Sometimes he needed truck lights to finish a job.

Most men understand this. When a wife needs to make a point of disgruntlement, she will say, "You are just like your dad!" That can mean a lot of different

Silky Fox ridden here by Doug Milholland in reining competition. Silky was fast, intelligent, and perfectly coordinated. In this sliding stop, he held continuous parallel tracks for 52', going from a dead run to a standstill. He was equally skilled at calf roping, steer roping, pleasure, and trail.

things; but most of the time for me, it was something about working all night and not coming home, being late for church, or putting the ranch and stock care ahead of the family. I was normally guilty as charged. Being like my dad was not always a bad thing, I thought.

Dad always had advice, which was rooted deep back in Depression days. "Son," he said, "always invest in land and cattle. The government can't print any more land. You buy one cow, she will have a calf in the spring, and then you have two." Dad referred to people who were "educated-fools, people educated beyond their mentality." He was a common-sense person. When I decided to go to college, he said, "Son, pay your own way. It will help you remember what you learn." I did, and it does.

Year after year, Silky paid the bills. Dad showed his foals and sold them to good, competent people. It is one thing to sell a great young prospect for a lot of money, but far better to sell one to a competent exhibitor who won't waste superior genetic talent. Dad was able to place a mare named Silk Worm with the legendary horsewoman and "Rose Garden" country-music favorite Lynn Anderson. She took Silk Worm to the largest shows in the nation and racked up 134 AQHA halter and 558 performance points. A top horsewoman on the Pacific Coast is Cynthia Cantleberry, so Dad sold some of Silky Fox's pretty mares for her professional hands. Cynthia won Reserve AQHA Super Horse and 158 Performance points with Silk Classic. Dad kept Silkay a few years and showed her himself. She won the *All American Quarter Horse Congress* in Columbus, OH, the *National Western Stock Show* in Denver, CO, and was AQHA World Champion. She earned 163 halter and 10 performance points.

All total, Silky Fox, to date, sired 433 registered foals and 80 performance-point winners, with a sum of 5,254 total AQHA points for all get. It is beyond question his progeny had looks and talent in abundance. They not only were great, they looked great in the presence of other great horses.

Silky Fox bred 21 full seasons. He enjoyed one owner from the minute the gavel dropped with Dad's bid. Silky had an enjoyable life. His paddock was a smooth-pipe pen right outside Dad's back door. He often stood in the corner of the pen and looked toward Pikes Peak (28 miles to the west) as thousands of people, who knew who he was, looked at him as they passed by on Highway 94. His name was written large on the all-steel barn behind where he stood. He was a celebrity personality in the community. Once a letter was delivered through the postal service addressed, "Silky Fox, Calhan, Colorado."

Every morning, bright and early, Dad would go out to the barn, lead Silky through the back barn door, and turn him out into a large pasture for a morning romp. Silky would tear out with the grace of his illustrious blue-blood ancestry, run a half-mile or so, do a few turns and twists, eat some blue grama grass, then, when he decided, walk back to the barn, where an open door and grain always awaited his return. This was the morning ritual, every morning.

On the morning of March 7, 1989, everything was the same. The breeding season was just starting, and two mares would arrive from Kansas for Silky's court during the day. Other mares were already in the mare motel waiting for the exact timing to breed. The pasture where Silky did his morning run was a mile

Frank Dickinson proudly displays his favorite stallion, Silky Fox.
Gary Lake photo

to the east fence. A neighbor called Dad and said there was a horse in the pasture trying to get free from wire wrapped around his legs. Dad knew only Silky was in that pasture. Mom called me to take a halter and go check on Dad; she had a bad case of women's intuition about this. When Dad arrived, no wire was apparent from a distance. As he got closer, his heart sank. It was a predicament far worse than wire. Silky had snapped the main bone in his hip and was in a dilemma he had never been trained to deal with. His reaction was to kick against the pain, and that was not going to help. We put a halter on him and tried to hold him still as he suffered the worst pain of his 24 years.

Our good veterinarian who did health work for the ranch was on call in the area and arrived in less than 20 minutes. A crowd had gathered in the pasture while a celebrity stallion was in his last crisis. In seconds, the vet said this was not going to be repaired. Silky couldn't be saved. The only solution was to put him down and eliminate the pain promptly. I told Dad that we would take care of Silky, and for him to leave—just go to the house, and he promptly did.

Dad slowly walked toward his pickup parked on the road side of the pasture. He turned and looked back after crossing the fence, as Silky watched him leave. They had been a team. Dad had lived with Silky, hauled him to shows in several states, watched him the first thing in the morning and the last thing at night, sat on the hay rack, and talked to him for hours. As Dad looked back for the last time, there was nothing he could do for Silky anymore. Each watched the other as Dad drove away.

The vet gave a full lethal dose in the vein, and within minutes Silky was gone. He just slowly lay down and that was that. The gathered crowd of well-wishers slowly departed. Chris Christensen was a friend of Dad's and an early arriver who lived the next ranch east. He graciously offered to get a large tractor with a front-end bucket and bury Silky. Within a half-hour, there was a pile of fresh dirt in the pasture and it was finished—nearly.

Mom had predicted that the day Silky died, Dad would die. Dad knew this event would be difficult, and his courage to be strong was less than zip. He also feared the personal trauma and his own ability to get on past this event. It was the day of the weekly Calhan Cattle Auction, so all felt it best for Dad to go to the sale, visit friends, and get away. He did.

Everyone wanted to talk about Silky. For weeks people came by to console Dad, but it wasn't easy.

My sister Vicky lived in Washington state. She called and was told the whole story, but she didn't feel right about the process or placement of the grave. She felt Silky should be buried in his own pen where he had enjoyed life for over 20 years. Gary Lake, our cattle foreman, felt the Christensen burial was proper, as this was where Silky's mares spent the summer. It was his big exercise pasture. Everyone had an opinion. Dad was ready to let a dead horse lie. Mom said it had already been sad enough and the burial was okay as is. I agreed with Mom, because Dad was not feeling good about exhuming Silky and doing a transplant. The emotion was such that my solution was to stay out of the discussion. Everyone had a special love for Silky, but Dad was, thankfully, still alive nearly two weeks after the passing, and no one wanted to stress him one bit more.

Tex Allen was a neighbor and one of Dad's many good friends. He had been gone on a trip, but when he heard about Silky, he came by to pay his respects. As the discussion of exhuming or not exhuming was massaged, Tex was ready to get in on the project. Mom wanted what was best for Dad, but she was against any additional stress. Most just wanted life to peacefully go forward. Everyone hoped Dad would heal with enough time.

There was discussion of how Silky was lowered into the grave. Chris swore it was in a careful respectful way. Others didn't agree. Vicky wanted him under his own shade tree, in his own pen. Then Tex decided the best and most respectful thing was to bury him in a standing position and have him pointed "looking" toward Pikes Peak, as was his custom for much of the last 20-plus years. The Pikes Peak direction was well received, but this was a long, tall horse.

The idea was to dig Silky up. Tex had a backhoe and a flatbed trailer and said he would do it for Dad, and for Silky, right now. Mom said no, she did not want Dad to see or worry about it at all. Tex came up with a plan for Dad to just leave, get in his truck, and attend the weekly cattle sale. It was two weeks to the day after the first burial. Go to the sale and don't worry about anything. Dad, as he was shuffled around by his closest friends, had somewhat of a good feeling about this bad situation, so he drove away. He wasn't thinking very straight and just wanted the appropriate thing for Silky.

A huge backhoe soon rolled down the road. I could handle it better than Dad, but I didn't watch. I was told Silky soon came down the road, horizontal on the flatbed trailer with over two dozen ranch vehicles behind him like a full-blown funeral procession. I never heard if all their lights were on or not. Silky was so tall and long it took a 12'-deep hole to get him covered in a standing position—with his head pointed up toward Pikes Peak. When Dad came home, allowing plenty of extra time for the second burial, there was another pile of fresh dirt and Silky successfully had his last burial.

We had a summer Texas Longhorn pasture south of Hartsel, CO, where there were huge rocks. We took a tractor up Ute Pass and loaded a large memorial stone of Dad's choice for Silky. Vicky had a special brass plate to be attached to the stone and engraved with "SILKY FOX, ONE OF THE GREATS." After Dad thought about it, he decided the plate was so nice he would just keep it in his house.

After Silky died on March 7, 1989. Dad bought, raised, and sold a lot more top horses, mostly related to Silky.

Frank Dickinson passed away on August 23, 2005. His obituary was printed in the *Quarter Horse Journal*, the *Denver Post*, *Gazette Telegraph*, and about 30 other publications. Under loved ones who had gone before, his obituary included "… and Silky Fox, AQHA Champion, who preceded him in death."   D‑D

Silky Fox and nameless rooster—oil on canvas titled "Continual Oat Shortage." This painting was selected for a *Western Horseman* cover.

"Whatsoever thy hand findeth to do, do it with thy might;"
∽ **Ecclesiastes 9:10**

How Not to Attend the AQHA Convention

This photo is of the Executive Board of the American Quarter Horse Association (AQHA) convention in the '60s. You will recognize major players: Clarence Scharbauer, Ed Honnen, Jack Schwabacher, Jack Casement, Lou Tuck, Bud Warren, Walter Merrick, Hugh Huntley, Howard K. Linger, and all the future powers-to-be.

I used to attend every AQHA convention to shake hands and meet these important horse people who would order paintings and photo sessions of their great horses. It hurt to miss this convention—bad. One time, when I had been married just a few years, at home in our little 8'x47' mobile home, kids were crying and there was no travel money.

That year, the AQHA convention was in Phoenix and I was broke. I couldn't afford 35¢-per-gallon gas to go to Phoenix nor the big downtown hotel room for

This 20"x 24" oil was sold as a cover for the *Quarter Horse Journal*. The original is in the Ed Honnen Estate private collection.

Linda and me for $45 per night. Still, I thought I had to communicate with this crowd. I needed these contacts. There I sat at home, discouraged. Every day of the convention, I made phone calls to the convention hotel and had them page the lobby for "Darol Dickinson." Every three hours, I would call the hotel "person to person." I didn't have to pay for the calls unless Darol Dickinson came to the phone in Phoenix, which he was not going to answer. All during the convention, I paged myself. I would ask that Darol Dickinson be paged in the ballroom, lobby, restaurant, everywhere the operator could find a paging microphone. Although I was never there, later several clients said they knew that in fact I was at Phoenix, because everyone was trying to find me—they said people were paging me all during the convention. Everyone, the important people, were hunting me.

Now, I realize that was devious, cheap, low-budget, deceptive, classless, but in those early years I badly needed name recognition. That was the total promotion budget I could afford—zero. ▷-◁

> *"Success is going from failure to failure without loss of enthusiasm."*
> ~ **Winston Churchill**

Vessels and Go Man Go

At our home in Colorado, snow storms were still very possible in April. For many years, I left home in my old car and drove to California to work. In mid-April, fine horses in southern California were slick and pretty. As slick hair and summer came later on further north, I zigzagged from southern to northern California, taking about a month every year to photograph many of the great horses in California. During that period, California had most of the great stallions and attracted most of the famous mares for breeding.

Many years before I worked up the California coast, in 1920, a young man, Frank Vessels, Sr., went west from Kentucky to the Orange County area. He started a small construction company and soon became astute at building oil drilling platforms out in the Pacific. He acquired 435 acres just south of Los Angeles to establish his ranch. Race horses were his passion, and genetics of the fastest stallions were his dream. In 1947, Mr. Vessels started holding local match races on part of his property that grew into the fabled Los Alamitos Race Course, the finest in the nation. Starting with local horsemen placing private bets, the sport of horse racing has grown to a multi-million dollar monthly industry thanks to Frank, Sr., and later Frank, Jr., who I worked with.

On my annual spring work in California (1960–1974), I could always expect photo jobs at Vessels' Stallion Station. The Vessels were leaders on the coast, owning Los Alamitos Race Course in the middle of everything. Chief Johnson was my connection at Vessels. I think he had naval military honors, plus he always had funny things to say. He held court in the office complex which was up in the Los Alamitos grandstand. He booked the mares and did horse deals. Over the years, I photographed all the top Vessels horses. Go Man Go was the great one in his era, a world-champion runner and an athlete of athletes.

How do you photograph a great stallion like Go Man Go? Do you hide his feet in grass? I decided to just shoot him naked, so we led him out of his stall and took him into the middle of the racetrack parking lot. He was what he was. He was

Go Man Go

not a show horse, but every horseman could study Go Man Go and learn a lot from the story of this runner's anatomy. He had some pretty red-roan coloring. He was quiet, gentle, and as easy to handle as any horse you could name.

Horsemen want a photographer to be honest. The camera elevation was exactly in the middle of his body. No retouching, no Photoshop. The Vessels used this photo in their *Pacific Coast Quarter Horse Magazine* ads. Shortly thereafter, Go Man Go was sold and went to Oklahoma where new, retouched photos were made. Never was an un-retouched photo of the horse ever published again, until now. My photos of great horses were never retouched. I worked hard at getting the best light, pose, and angle—and that was the naked truth. True horsemen appreciate honest photos. The deception of Photoshop horse photos today is unbelievably extreme. Every good horseman can see the fakey stuff.

> *"Seest thou a man diligent in his business? He shall stand before kings; he shall not stand before obscure men."*
> ~ **Proverbs 22:29**

Ruidoso Downs, New Mexico

Ruidoso, high in the Rocky Mountains of southeastern New Mexico, is Billy the Kid country. He came to the area to hide out in the mountains, and today Texas people come to the same area in the summer to hide out from the Texas heat. It is a beautiful pine-scented area that provides the aroma, food, games, and the gambling spirit for horsemen from all over.

The All American Futurity is a 440-yard race for 2-year-olds run every Labor Day. It first grabbed world-wide fame when the prize money exceeded that of the Kentucky Derby. Now the estimated purse is over $3,000,000. Obviously, that net draws in a lot of excited people to see it all happen.

Late in the summers, I migrated to Ruidoso, where I attended the horse sales, sat around the stables, met owners, jockeys, trainers, and spectators. The club room high up in the grandstand was a watering hole for the visiting high-rollers, but somehow I managed to get in. I rubbed shoulders with famous people of the day like actor Robert Mitchem, Dale Robertson, John T.L. Jones, Jr., Buddy Preston, Hugh Huntley, D. Wayne Lukas, Walt Wiggins, Frank Vessels, Howard K. Linger, J.B. Ferguson, Grafton Moore, and some brightly dressed ladies who were not always on their best behavior. It was a place where people were excited and might order oil portraits of their hopefully famous, or already famous, horses.

In 1968, I had two commissions to research during my stay at Ruidoso—the World Champion Runners Bar None Doll and Brad Lin. The owner of Bar None Doll wanted her immortalized in full race-tack posed in front of the Ruidoso Downs tote board. He didn't want the grandstand, just the landscaping, small rolling hills, and race track for a background. Never one to make things easy, I planned in detail, from the Kentucky Bluegrass with dandelions obnoxiously demanding the sun, to the specific flowers and carefully manicured evergreen bushes. I did sketches for the owner to give an idea of how the finished painting would look. I spent time with the mare, her trainer, and owner. A portrait artist has to prepare for criticism from the veterinarian, the farrier, the trainer, the guy who rubs legs and brushes her, and the owner who has spent months or years carefully enjoying her beautiful anatomy. If something is not correct, someone will complain.

World Champion Bar None Doll—oil on canvas 30"x36"

In western art, the easiest thing to paint is clouds. They can be any shape or color. The hardest things are horse anatomy, horse feet, and people's faces. When a painting has a cowboy's face turned away, it tells you that the artist doesn't want to, or can't paint a face. When a horse's feet are intentionally hidden in dirt or grass by the artist, that also tells you something about the artist. Early on, I worked at doing pencil drawings of faces and horse feet. It was an exercise I had to force myself to learn. Either I faced the challenge or had to give up early.

The Brad Lin oil was a commission from ground level, Ruidoso Downs grandstand view, looking east. Owner Buddy Preston said Brad Lin had a certain look coming onto the track when he was ready to win. If Buddy saw that look, he bet a lot of money. He wanted that look in the painting. It was an alert look, high head, spirited attitude—somewhat wrestling with the jockey and a sound spring to his step. I went to several races to watch Brad Lin for this look. I did sketches and photos of jockey Richard Bickel during his waiting time in several different tracks' jockey rooms. Most things in art are just a matter of serious planning, then detailed execution. Artists can't ever get in a hurry when doing their best work.

During the early '60s up until the late '70s, I completed 10–14 large oils per year. When doing *The Color of Horses* book, I took about 10 days each for 34 different colors. My normal backlog was over a year. That project was a big piece of my life. It fed the family, made some land payments, and bought Texas Longhorn cattle—that was all very good.   ▷-ᴅ

World Champion Brad Lin TAAA—oil on canvas 30"x36"

I first met Robert and Dorothy Mitchem in the club house at Ruidoso. Dorothy could never find something Robert wanted for Christmas, because he had everything he ever dreamed of, so she would order an oil portrait of one of his many horses. Belmont Scare was a AAA son of Spanish Fort, which he purchased from Dale Robertson of *Wells Fargo* TV fame. The background is in the Santa Ynez, CA, valley where Robert kept his stable of running horses.

The Diamond Charge painting was commissioned by R.D. Hubbard, owner of Ruidoso Downs. Diamond Charge was the first horse to achieve AAA+ at all four recognized track distances. The background was north of Wichita, KS, at the Red Bee Ranch owned in partnership with Art Lankin

Ruidoso Downs, New Mexico

The Johnny Cash Longhorn Herd

Lawton, OK, at the Holiday Inn on Cache Road, was the center of the Texas Longhorn world for many years. There was one annual sale, at the Wichita Wildlife Refuge, and one annual get-together called the Texas Longhorn Breeders Association of America (TLBAA) Convention. The whole thing was a 2-day deal. If you attended during the mid-1970s, that was it for the whole year. Every serious Longhorn person was there, and it wasn't hard to shake hands and visit with all of them. They were a close-knit group.

We first met T.W. "Wick" Comer at the Holiday Inn restaurant at breakfast. He and his lady friend Big Red were eating breakfast, and no one acted like they noticed them. With a few introductory pleasantries, we found them very likable people and excited about attending their first Texas Longhorn sale. Over the next several years, Wick attended every Longhorn function. Whatever their other business was, they escaped and did not miss a single event. Wick had a happy personality. He always had a friendly smile like the wave on a slop bucket, but the men all swarmed around Big Red.

Big Red was not heavy, but trim in type and perhaps 5'11". She had flaming red hair, talked loud, and was easily the life of the party within minutes of her entrance. She dressed with flash and tease. The women would corner her and talk "shop." Big Red promptly told the girls she wasn't married to Wick; she worked for him on trips and weekends. When he traveled, she went everywhere with him—but she had her own business and wasn't going to marry him. She worked as a "décor-employee." Big Red attracted people, and Wick was right by her side. When he overdid the liquids, her job was to manhandle him in her own way, load him up, and get him settled in the motel. She did not drink on the job. She was proud that her 24-hour fee was $500. Big Red would tell the girls everything she did; all they had to do was ask.

Wick's dad, Guy Comer, according to the *Nashville Post*, built Washington Manufacturing. By the late '60s, the company had over $250 million in revenue and up to 10,000 employees in its many plant divisions. The apparel industry in the South was as solid as the history of cotton farming. Guy turned operations over to Wick at a time when Wal-Mart was sinking its teeth into every mom-and-pop business in the nation. As garments started flowing into the U.S.A. by the boatloads from China, Wick had his hands full to hold the company together.

Wick had previously been in the registered Red Angus business and owned several of the leading sires of the "Chief" line. The registered business was not new to him. He had learned that the very best ones were the most profitable. If you weren't using bloodlines at the highest level, business was not as profitable, nor as much fun.

Wick inherited huge responsibilities and had huge goals. His Texas Longhorns and many friendships provided him an enjoyable life in the beautiful hills of Hendersonville, TN.

In 1978, the prettiest black-spotted bull was born out of the great Dickinson Cattle Company (DCC) cow Ranger's Measles (also dam of Emperor). He was named Impressive—and he really was. He had a beef type with a nice full hip and straight back, and you could pick him out a mile away in the herd. He was sharp. We had planned to breed him for several years, had used him as a yearling and a 2-year-old—then Wick heard about him. Wick wanted a price, yet we were

Wick lived in a beautiful part of Tennessee. He liked Paint horses, and his favorite bull was Impressive.

not planning to sell, no way, no how. However, for some reason I finally priced him for the odd number of $62,000. Wick seemed to have no pain with that number, and in a few seconds he owned Impressive.

In the early days of the breed, most cattle were small. We worked to get larger size and correct type in the herd and were starting to get a few cows weighing over 1,200 lbs. and a few with horns over 50". Wick was hunting over-50" cows weighing more than 1,200 lbs. He heard about our cow Ghost, who was well above both marks. She was getting some age, and we had several calves from her by embryo transfer. Wick wanted Ghost and offered $20,000, so she left along with some of her daughters. Wick was good to us in every way. He and Big Red were a fun-loving and flamboyant pair.

Then something happened, and Big Red was replaced with a lady named Lynda. She came with Wick to some cattle events and seemed like a nice enough person. Wick announced that he was going to finally tie the knot, although it was hard to believe he would settle down with one woman. He and Lynda were going to get married right in the cattle-sales arena. And of all things, the ceremony would be held just before a Texas Longhorn sale. Wick spoke highly of Lynda and the wedding was, according to them both, really going to happen.

The sale came and went, but still no photos or any word of the wedding. Wick's report was, "At the last minute I found out her true nature, what kind of a person she really was, and just in the nick of time." He had an old mosey-faced mutt dog that went everywhere with him on the ranch. He was always trying to catch a Canadian honker or dig up the flower beds. He had a personality that would be easy to do without, but not for Wick. He slept with Wick. At all hours of the night, Wick would let him out for a relief drain. Wick continued, "We were getting into bed the week before the wedding and my dog crawled into the bed. He always slept right beside me. Well, she didn't like that. She said to get the dog out of the bed! There was no settling her down. Finally I told her the dog owned the house and he was just letting us use his bed. Then she drew a line in the sand and said, it was her or the dog. At first it bothered me a lot, then I felt relieved to see her go."

A few months after the wedding folded, Wick decided to have a big Texas Longhorn auction. It was something he had done before, and he wanted a full-color catalog, good photos of all his animals, and some colorful scenic shots for advertisement decorations. I had never been to his ranch, so I offered to fly to Nashville and spend a couple days taking photos for him. He was a good client, and I wanted to do the best job for him that I could to make his sale top-notch.

Wick had a beautiful ranch behind a rustic security gate near Hendersonville, TN. His ranch was rolling hills, grass, acres covered with wild Canadian geese, huge trees, a large lake, and a rustic but elegant home. He owned the main bank in Hendersonville where the Oak Ridge Boys and several other entertainers banked.

Wick saddled up a pretty ranch horse and personally drove the herd over hills, under overhanging trees, and around his large lake. It was easy to get colorful shots for his project. Impressive was matured, so I got new photos of him showing his development.

We went out to a good steak place, and Wick shared his heart with me. He had been a fan of Johnny Cash forever. Johnny lived very private, near Hendersonville, and Wick had tried to contact him, but his reclusive attitude had shielded the "Man in Black" from the public. Wick decided if he could get Johnny into the Texas Longhorn business, the promotion would be outstanding. At this time,

Johnny Cash was all over the country music thing like a fat man on a small bar stool. Wick went over some things he had done to connect with Cash, but nothing worked. He would give anything to get Cash raising a herd of Longhorns on his sprawling Tennessee estate. He knew, with the world-wide recognition of Johnny and his wife, June, interest in the breed would explode. That would be real professional marketing. If the Cash family would come to Wick's Longhorn sale, attendance would jump a thousand people. It was a noble goal, yet those high-dollar entertainers, like a goose in a hail storm, wake up in a different world every day. They are definitely a moving target—but possibly worth it.

When we were finished, Wick took me to the airport. On the way from Hendersonville to Nashville, I saw several unused cow pastures from 20 to 200 acres right along the highway. Thinking out loud, I asked Wick if Johnny Cash traveled this road. He knew for a fact the "Folsom Prisoner" drove to his Nashville recording studio nearly every morning about 9:00 right down this same road.

If Wick was serious, and he said he was, I recommended he lease one of the farms along Tennessee Highway 31. Put some pretty Texas Longhorn cows along the road. Get a colorful sign saying TEXAS LONGHORNS FOR SALE, clearly visible from the road. Wick had a cowboy named Jimmy, so I suggested that

Wick's rustic home located on the north slope of Old Hickory Lake

Jimmy go out about 8:30 AM and put cattle feed right inside the fence along the highway. Have those pretty horned cows right there so Cash could not miss seeing them morning after morning. Wick just smiled and didn't comment.

He called about a month later and was really excited on the phone. He had found an old friend who owned one of those farms along the road and had a sign up. Jimmy was feeding every morning and thought he saw Johnny Cash go by last week in a black car. It is exciting when a plan falls in place.

Several months later, I got a call from Mary Beth Vineyard, who lived near Houston, TX. I had sold her some Texas Longhorns, and she was having fun raising fancy cattle. She was excited. "You won't believe what happened," she said. "Johnny Cash has a business manager in Houston. The fellow called me and said Johnny Cash wants to buy a 'real' Texas Longhorn from Texas and wants to have someone deliver him to Hendersonville, Tennessee. Can you believe that?" As a matter of fact, I could believe it.

Wick saw visions of dollars from the Country Music Legend for the purchase of a full-blown herd. Another Longhorn guy and neighbor of Mary Beth's named Crawford Boyd said no problem; he would just give Johnny Cash a big steer at no charge. There was TV and newspaper coverage on the Houston end. A lot of press was attached to the departing steer that was soon to be a *bona fide* Texas ambassador to Tennessee. It was a fun thing that just dropped out of the air for no reason.

Mary Beth wanted to meet Johnny Cash and make sure everything went perfect for the delivery and the presentation. She knew the steer would be tired and probably get some steer mud splattered around. For a smooth coordination, she needed a place to rest him a day before making the presentation. To make a good impression, she wanted to rest the steer, give him a bath, and get him full of hay to look his ultimate best. There was only one good ranch for the rest and relaxation—Wick's place.

I gave Mary Beth the phone number for Wick. He was a gentleman in every way and kind enough to help her out. She flew to Nashville while Crawford's son, Frank, and Mike McLeod hauled the big steer to Nashville and on to Hendersonville. After the rest stop, he was delivered to a really fancy estate where Johnny

Cash, June Carter Cash, and their son, John Carter Cash, lived. The welcoming family drove out to the pasture in a shiny black Mercedes to greet the arriving steer. Mike McLeod reported, "They were all smiles. The Man in Black threw us a curve and was dressed in a white T-shirt and white shorts. There were lots of photos taken of the dressed-down celebrity. The pictures came out in several papers and the Longhorn magazine. We all stood around and talked for a while, then just loaded up and drove back to Texas."

Johnny Cash never attended Wick's sale, and no one has ever seen the steer since he was delivered. Why pay for cattle if someone will donate them for free?

I hate it! I really hate to end this account this way, but that was what happened. So much for my bright idea of getting Johnny Cash into the Longhorn business and creating the Johnny Cash Longhorn herd.

"I've learned that people will forget what you said, people will forget what you did, but people will never forget how you made them feel."

~ **Poet Maya Angelou**

Appaloosas and Carl Miles

Carl Miles was an oil man, a wheeler-and-dealer of the highest order. He once told me he bought over 400 dry oil wells that had stopped profitable production. I didn't understand it, but through some process he was aware of, he opened them up and got most back into profitable production. With the well purchases, he got mineral rights and later sold them for deeper-well production. There was more oil by far in the lower strata.

Carl Miles was a friend of Cecil Dobbin, who owned Bright Eyes Brother. Cecil told Carl that he should get me to do portraits of his top stallions. Over the years, I painted one of Joker B, Chicos Snowcap, Mr. J.B., and Prince Plaudit. Each time it was an experience to talk to Carl. In minutes, you could learn more than a semester in college—Carl had been there, and did it. He was a bold risk-taker—a charming fellow, friendly, always with a Dwight Eisenhower smile.

This photo of Carl was taken at his place south of Abilene, TX. I was doing the research on Prince Plaudit for an oil portrait, and I asked Carl to bend over so Prince Plaudit's beautiful hide would be the photo background. Someone watching was wearing an economical straw hat which I borrowed for Carl. Harry Reed

Carl Miles

Elmo Favor, AQHA judge, dear friend, and Prince Plaudit ear-getter-upper.

(the horse trainer, not Senator Reid, D-NV) was holding Prince Plaudit, and Elmo Favor was helping get ears up. There is another photo in the ***Photographing Livestock*** book that was taken the same day, during the photoshoot with Prince Plaudit. It also shows some special photo-preparation techniques.

Hank Wiescamp dabbled in Paint and Appaloosa horses. He raised some of the best in each breed, but his heart was in Quarter Horses. All his horses were basically the same bloodlines—linebred Old Fred. When Carl heard about Prince Plaudit, he immediately started working on a deal with Wiescamp. No one could smell money like Wiescamp, so a fellow like Carl was a choice guest. Wiescamp could smell his wallet when he boarded the plane in Texas.

Carl went to Alamosa, CO, and bought Prince Plaudit; but while he was there, Wiescamp teased him with a drive-by look at over 400 mares, mostly of Skipper W breeding. Carl asked how much for that one, and how much for that one over there. Hank refused to cough up any prices. He just kept driving his big Chrystler Imperial across the rough irrigation ridges, talking and joking about everything from King David forward. Carl went back to Texas and continued to call daily about buying mares. Finally, Hank agreed to put some packages together for Carl to consider. He was priming a big pump.

The day came. Hank took Carl to a pasture showing him 50 mares which, according to Hank, would work well on Prince Plaudit. He reportedly priced them for $5,000 each. In an hour, they drove through the mares and Carl said okay—he would take the package. He then asked if there were more, and Hank said he did "just happen" to have another pasture nearby with 50 mares, but they were the top end. Carl could take them to Texas for $10,000 each. He took them all. That was the foundation of the Prince Plaudit breeding program, which dominated the Appaloosa world for years.

Carl Miles was a cunning promoter. Every cattle or horse breed needs one really brilliant promoter who will cut the first machete trails through the jungle. True, he may get the most fruit as he goes, he may fall into a sink-hole, but he cuts the trail for the rest of the world to safely follow. A breed association doesn't necessarily need two leaders, but every breed that has had great success and created great value absolutely had to have at least one promoter. Without one serious leader who really savvies breed promotion, no breed has ever excelled.

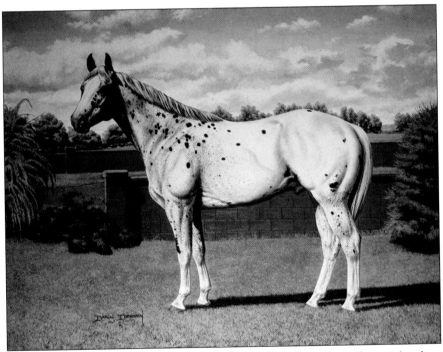

Oil painting of Prince Plaudit commissioned by Carl Miles. The setting is the front yard of Miami Oil Company south of Abilene, TX. Carl loved this stallion and watched my photo efforts with great interest.

This shot was taken of Prince Plaudit behind Carl's office south of Abilene, TX. Prince Plaudit sold in the dispersal sale for $260,000.

Someone has to carry the big buckets of water—and for Appaloosa horses, that important giant was Carl Miles.

After years of watching Carl's brilliant promotion of Appaloosa horses, I once asked him what his secret was to move so fast and be so successful so quickly. This was his answer, which I will never forget: "I am a diabetic. Diabetics don't live as long as other people. I always figure I have 1 year left to live. I plan every investment to culminate in 12 months. I can't make the mistakes other healthier people make by giving themselves too much time."

Carl passed away in 1976 while flying over Pennsylvania in Dave Stahlman's private plane. Prince Plaudit died in 1988 at 25 years of age. A lot of legacies were made and millions of dollars of horses evolved from the Carl Miles program.

"A game changer is that ah-ha moment where you see something others don't. It's the transformational magic that takes one from the ordinary to exceptional."

~ Mike Myatt, *Forbes Magazine*

Thanks to referrals by Carl Miles and Cecil Dobbin, many of my earlier commissions were of Appaloosa horses. This is Bright Banner by Bright Eyes Brother, a 30"x36" oil for Chuck Singleton of Grass Valley, CA.

Bogus Bonus

Bogus Bonus—In 1998 a beautiful cow, Quilla Fet, was sold to Ralph Demshar of Ravenna, OH. Ralph had a high-tech camera store, did a world of business near the university, and decided to start raising Texas Longhorns. He had a pretty place in a beautiful Ohio semi-rural area where he raised blooded Arabian horses. He wanted a few Texas Longhorns, but only the very best. Quilla Fet was the best Quill daughter ever. She was huge, dark seal-brown, spotted, and bred to The Shadow. At that time she was pretty pricey, but Ralph bought her anyway. The following year, he sent a photo of one of the prettiest Shadow calves ever. He was spotted like his dam, with a huge hip and everything you could want on a young bull. I asked about buying him—the answer was no. Ralph had three cows, including the bull's mother; and when he was old enough, he wanted to use him for breeding.

Every year or so, Ralph would send a photo and I would try to buy this unregistered bull. It was starting to appear that he was the best son of The Shadow, and he just kept getting better. I asked to buy him nearly every year as time went on. Always not for sale. Ralph called him "Grump."

When he was about 7 years old, something happened. I didn't get in the loop. Ralph sold his little breeding package of Grump and mostly his daughters to a neighbor. The new owner sent me photos—this bull was really nice now. Again not for sale. Every owner teased me with photos of this special bull. He was so gentle. He would eat out of your hand, but he was a little pushy. If the feed didn't come quickly, he would slowly, methodically just walk over whoever was holding the bucket. The owner got a little scared of him. After a few aggressive moves by Grump, the owner got real scared. To increase the problem, over the years some bull calves had been born that were growing into bigger bulls. Now they were old enough to fight Grump and were regularly tearing up corrals and fences. The owner called me in the fall of 2008 and said he had decided to shoot all the cattle as he was scared of them and couldn't load them into a trailer. He

had tried. I made an offer of $1,000 each but had to take delivery in the pasture, with no promise of capture or corralling of any kind. Grump was finally mine—if we could catch him. I was excited. A great new bull is always exciting.

I called Denny Thorsell from Medina, OH, who owned the Bucking Ohio Bull Riding Arena. Denny's son Shawn is a regular bucking-bull handler. Bulls a lot meaner than Grump were a walk in the park for the Thorsells. I employed Shawn and some really serious cowboy friends to go capture Grump and his family. It wasn't easy, it wasn't pretty, horses were sweaty, some fences were wrecked, and the owner was scared to death along with some big-eyed neighbors. However, about four hours later, the cattle were unloaded at DCC, tired, scratched up, battered, yet alive.

We ground all the bulls except Grump. He was beautiful. He was never a problem to handle. We had immediately decided to collect semen on him, but the plan was to wait until after breeding season. We registered Grump as "Bogus Bonus." The name came at the time the federales in charge of Fannie Mae and Freddie Mac had fraudulently used the taxpayers' billions and given themselves a big bonus.

The Shadow was a breed-improving bull. Three of his progeny that made a special contribution include Shadowizm, who sired Jamakizm, Mile Marker, and the clone cow Shadow Jubilee ($120,000), one of the first over-90" T2T cows. Bogus Bonus was one year younger than Shadowizm and a real genetic find at the time. Up until this special breeding season, his calves had all been from his sisters and his mother.

When the day came, in June Bogus was given about 50 cows. He was so happy. You could see the smile in his heart. If a bull could do cartwheels, he would have been doing them. As I checked his pasture, he was on the move—constantly checking for new business, just being friendly. He would even break into a trot between cows. He did not want to miss any personal opportunity. His life had changed from 4 cows, which included his sister, his sons, and his mother, to 50 unrelated cows, and he was happy, happy, happy!

About a week into breeding season, I saw Bogus standing in Muskrat Creek. It was hot, so that was not unusual. I told the ranch cowboys to check on him, and the next day they said he was still standing in the creek. Finally, after about three days, I said drive him out of the creek up to the cows—then they saw it. His leg was in terrible shape. It appeared that somehow he had entrapped himself some way, cutting the hide right below his knee so that all his skin had slid down around his foot. The bone and tendons between his knee and ankle were mud-covered and totally exposed—contaminated in every way, swollen up huge.

On July 10, 2009, the vet said there was no way to save him, and he was right. For people who have never dreamed of raising great cattle, superior genetics, or the very élite in bloodlines, perhaps this event is of a minor emotion, but that was not the case with our family. It felt like we had all been whacked in the head with a bodark post.

Most ranchers don't hibernate and cry for days when a valuable critter passes, but within your gut a deep hurt lives that doesn't go away, maybe for years. There are no crisis-therapy classes to attend. It doesn't help to take off work. The best recovery is just to go back to work and get positive, productive accomplishments done. Try to think about other good things.

His skull was polished into the most rugged, beautiful western display.

The following year, 5 calves were born from Bogus Bonus. No semen was ever collected. A beautiful dream turned into a sad nightmare. Sorry, but that was the Bogus Bonus ending, about like Fannie Mae and Freddie Mac.

"Everyone can master a grief but he that has it"
~ **William Shakespeare**

The Rest of the Butler Story

Starting in the mid-1960s, I traveled the nation trying to learn about Texas Longhorn cattle. In my original profession of doing oil-paintings of famous livestock, I had to research the subjects wherever they were. This often took my camera and me to Texas. During my expense-paid travels, I diverted to study every Texas Longhorn herd I could find. There were no magazines, weekly sales, or shows during those years. Studying cattle and genetics was totally about going to the herd, and still is today.

The Wright herd at Robstown, TX, was fascinating. Its origin went back to the early 1900s, when everything was multiple-sire herds with 20–30 bulls and 300–400 cows all running together on the Nueces River bottoms. Texas Longhorns were a Wright family thing of multiple generations. About 1972, I saw some of Wright's flat-horned cows that were standouts. When I was quizzing M.P. "Chico" Wright, he said these were Butler cows. At early (1964) Longhorn sales, Wrights bought Butler cattle because of their larger horn. Very interesting this gem of information. The Wrights would band their own beautiful, flat-horned, feminine cows with the sons of Butler cows.

The same year while I was visiting Travis Marks' Texas Longhorn herd at Barker, TX, he was using a bull with more horn than I was accustomed to seeing. He said it was a Butler bull, and one man still had a few of the remaining Butler family of cattle, F.M. Graves of Dayton, TX. That evening, my rental car arrived at Dayton. I had the enthusiasm of a dedicated coonhound on a promising hot trail. In my heart, I believed the Texas Longhorn breed could be brought out of the swamps of Texas and into great value if marketed and managed correctly. I wanted to make sure that was done.

F.M. Graves, whose nickname was Blackie, lived in Dayton and farmed rice all over the county. He showed me the largest-horned cattle I had ever seen. I tried to purchase his whole herd of about 15 cows. He refused, but eventually let me select 3 cows of my choice. He indicated something was complicated and some-

F.M. "Blackie" Graves, a man of his handshake, was regarded as the main breeder of Butler cattle. At the local Raywood auction, he had easy access to hundreds of Butler cattle, mostly selling for less than $350. The highest-selling cow he ever bred was Royal Reputation by Dixie River and out of Texas Trish FM 276, who was a daughter of Texas Ranger. She sold to Richard Carroll for $150,000 in the Legacy sale.

how Henry Butler, son of Milby Butler, was involved. I tried hard to purchase a red cow later to be named Miss Dayton #5. She was tied up in the legal complication, but I was able to purchase 2 of her daughters for $250 and $300 each.

Blackie wanted to show me a white bull with black ears that he was using on all the cows. The bull was purportedly in a 40-acre grass pasture. Blackie drove back and forth but could not find him. Although the grass was not abundant, there were patches of thick, tall weeds. Blackie was frustrated that in this small pasture he could not find the bull. There were cows all around. Finally, right in front of the truck, the white bull was lying with his head flat on the ground like a crouching coyote. When we drove closer, about 50' away, he jumped straight up like he was shot and stood there scared to death—frozen still. We looked him over. I took a photo. Blackie admitted the bull was a little spooky acting. He appeared to have the largest horn I had seen up to that date, although measurements by pure "guesstimation" from a distance aren't always accurate.

Although Blackie wasn't interested, I tried to buy him. He said no, he needed him one more breeding season and then in the fall I could come get him. We agreed on a price of $300. He said that was what he would bring at Raywood

Livestock Market. I kept in touch with Blackie during the year and looked forward with exhilaration to buying this spooky-eyed Butler bull. I waited for Blackie's call.

In late summer, I called Blackie about picking up the black-eared bull. My heart dropped. Even though we had agreed on the bull and talked about the purchase several times—he had sold the bull for slaughter! I was crushed. I felt like I had been beat over the head with a frozen skunk! Blackie said he had gathered the herd and removed the bull for me to take. When he came back the next day, the bull had ripped up a heavy board fence, crawled out, and was back with the cows. His men fixed the corral, made it much stronger, and gathered the herd again. The bull was contained and the cows turned out. This time the bull trashed a whole section of really stout fence and escaped. Blackie's men gathered him again, loaded him in a steel trailer, and sold him at the local Raywood auction.

I was depressed. I thought DCC had better facilities and could have handled him. We would have also collected semen and preserved this rare Butler bloodline. Blackie easily sensed my discouragement and offered to sell me any of the bull's sons. I selected a one-year-old white bull with black ears out of the cow I couldn't buy, Miss Dayton #5. The price was $250. We branded him with an ID number 250 on his right rib and registered him as Bold Ruler.

After selling the wild bull, Blackie used another Butler bull he had acquired from Buck Echols, the county sheriff. Blackie also promised to sell this bull when he was through with him, and he did. We registered him as Man o' War. Again the going price was $300.

Later, Blackie called, which was unusual. He said he had read an ad about a bull over by San Antonio who measured 52" tip-to-tip. The ad boasted the widest-horned Longhorn bull in the nation. Blackie said he had a bull that was about 58". That was really exciting, so I asked him for a recount. When he called back, he said that he had, in fact, been wrong. The bull was actually 59". This was for sure the widest-horned bull ever, well past all others. I could see big dollar signs. I asked what he wanted for him, and he said he was not for sale. The bull wasn't registered, so his calves couldn't ever be registered. Later I learned he had purchased him in the weekly Raywood cattle auction for $317 with no guarantees expressed or implied.

The Rest of the Butler Story

Blackie was pretty well-heeled, but most people do have a price. Edwin Dow of Fort Worth and I got on the phone and called a few people. Fairly quick, 10 other Longhorn producers agreed that if we could buy him, each would put in $1,000. That would be the first $10,000 Texas Longhorn in history. Edwin and I drove to Dayton through a bad storm and couldn't get a motel room. He and I shamed a motel desk clerk into allowing us to spend the night on a couple of old army cots in a janitor's room on the way to Dayton. It was hard to sleep for several reasons, mostly because it was near certain that the next day we would own the largest-horned bull in the world.

In Blackie's yard, Edwin and I loaded into his pickup. He moved papers and tools from the truck seat to make room. His dash was several inches deep with a random assortment of small tools, papers, and an Afrin squirter which he regularly applied to his nose. As Blackie cheerfully drove down a Texas farm-market road, he began to mutter loud and became unusually mad. He had noticed one of his newly planted rice fields covered with thousands of white tropical birds. He braked to a dead stop in the middle of the road and started feeling under Edwin's legs. Grabbing a high-powered rifle, he started blasting away all over the field. In seconds, the birds were all in flight. None remained. In defense of his aim, Blackie quickly told us he wasn't really trying to kill the birds; he just wanted to scare them away from eating his newly planted rice seed—over to a neighbor's field.

Blackie took us to the Longhorn pasture where by now he had about 40 cows with 3 adult bulls. There was the 59-incher, a red-spotted one, and a high-horned—beautifully speckled—multi-colored bull (later registered as Conquistador). The 59-incher was white and was being hammered by the two older bulls. The purchase was a no-brainer, so Edwin made the $10,000 offer without a twitch from Blackie. He turned the offer down flat and said, "No deal." Apparently, the big-horned, black-eared bull was not registered and never would be. Blackie had been using him anyhow—but registering his calves out of a deceased older bull named "Sam." His calves were all registered as sired by "Sam." In the industry, the Sams were referred to as "Real Sam" and "New Sam." Soon there might be a "White Sam" and maybe a "Speckled Sam," but deceased Sam was the only registered bull that Blackie had owned. The ethics of his plot were discouraging. It cost only $15 to register a Texas Longhorn at that time.

Blackie explained the legal battle. Milby Butler, who was deceased, had had a serious "friendship" with a Pauline Russell who lived near Liberty, TX. The elder Butler was divorced from his lifelong wife due to this "special" friendship. In his will, Mrs. Russell received 300 acres near Ames, TX, all the cow herd, and numerous things of value. Milby Butler's own family received only a few properties. The estate was further complicated by a divorce-filing from Mrs. Milby Butler just before he died. The court jostled from 1972 to 1978, splitting, liquidating, and dispensing assets. A royal battle ensued alleging inordinate favors Mrs. Russell had performed to get the lion's share of this sizable estate. Blackie had been a witness and testified in court, siding with the Butler family, so Mrs. Russell hated him. She had raised the 59" bull and was the only one in the world who could legally sign the papers to register him. Blackie believed there was no way in Texas that could ever happen. He knew exactly what Mrs. Russell thought of him personally, and to her death she would not change come hell or high water.

The legendary "Real Sam" was blamed for a lot of things. His siring abilities included calves from several pastures, all at the same time, during his life, and for years after his death. He was the 15th bull registered in TLBAA.

Having failed to purchase the 59" bull outright, I made an offer. I would make every effort to get the bull registered, totally legal, with correct paper work if he would give me a half-interest in the bull. We would be 50/50 partners. With nothing to lose for Blackie, it was a deal. I offered to draw up an agreement, but he said, among gentlemen there was nothing more binding than a friend's handshake, so a handshake it was.

The Rest of the Butler Story

Edwin and I drove away with heads spinning. This would not be easy. It might be impossible, but I wanted that bull really bad. At that time, the whole Texas Longhorn industry needed this bull.

As a result of my travels to view hundreds of cattle, a pattern was falling into place. I had thought there were 6 unrelated families of purebred old Texas Longhorns: Marks, Phillips, Yates, Wichita Refuge, Wright, Peeler, and now Butler. With the Butlers, there were definitely 7 families bred before 1930 that were reasonably unrelated. These 7 herds were bred for dozens of years in the early 1900s without mixing outside genetics. The Wichita Refuge was the last of the 7 to be a closed herd. The Butlers were the smallest in number and the least known. As I started to work at increasing this family in our breeding program, I wanted the history and special traits to be correctly recorded. A historical chronicle was needed to record this fascinating herd. People who love the cattle like to know this type of history, so I called everyone ever involved with the Butler cattle, including Mrs. Pauline Russell, trying to put the facts together.

"New Sam" was used by Blackie Graves for several seasons. He is the sire of Bold Ruler and many early-registered cattle.

When Milby Butler died, a sizable herd of cattle was transferred to Mrs. Russell. The fond connection with Mr. Butler was due to their mutual devotion and love for the lifetime task of breeding this family of special cattle. Mrs. Russell had the land and cattle to continue to love and breed the herd. Yet, within days of the reading of the will, Mrs. Russell hired cowboys with trucks, and the main herd went to Raywood auction. These cattle at Raywood reportedly had the most serpentine appendages in one pen that viewers had ever seen. Most cattle sold for weight prices and were hamburgered within two days. About a dozen area Longhorn aficionados bought certain select Butler individuals and preserved some of the best ones.

What didn't go to Raywood was a bull registered as Bevo and a cow registered as Beauty. Mrs. Russell and her husband, Wiley, retained these two in a small grass patch behind their house south of Liberty. From the early 1970s, these reproduced. It was a close family: Bevo was the sire of everything. Some believe he was a son of Beauty, but Mrs. Russell said that was not so: he "wouldn't service his own mother." By 1976, the Russell herd multiplied females and even more bulls. Again, Raywood auction got even more cattle, but Mrs. Russell kept Beauty, that she and Mr. Butler reportedly were very fond of. Beauty was the widest-horned cow in history at that time, with record corkscrew-twisted horns measuring over 60".

Mrs. Russell met each call I made to her with suspicion and fear. As I asked her about the history of the cattle, she suspected I represented some law firm on behalf of the Butler family. I called her about twice a month. Sometimes she was friendly, other times very defensive. I wanted to see the famous Beauty. I tried to buy her, yet nothing happened. I wanted to personally interview Mrs. Russell, but she was scared of me. I sent her photos of our cattle, my children, me, and anything I thought would start the first step of a friendship. She was cold and sometimes would just hang up the phone as I was talking. I would call again later, as if all was well, and keep trying.

Finally the day came. I called her and said I was going to be doing some business around Houston. I wanted to meet with her and do an interview for an article on the Butler cattle and see the registration certificates on her cows. She agreed. In those days you could go to the airport, buy a ticket, and board a plane a few minutes later, so I did.

I checked into an old plywood motel a few miles from her home and called her for an appointment. She said she had a black boy coming who was going to help her with flower beds, her yard, and work around the house. I agreed that was pretty important and she was doing the right thing, so I waited in the motel. The next day I called again. She was very busy and recommended I should come by another time when I was on business near Houston. Well, she was my business—my only business.

During the day, I visited with Booster Stevenson, DVM, who knew her. He said she was a little funny and he had no recommendations. Booster had bought one of her bulls in the last Raywood sale and certainly wished me well. He wanted his bull registered but couldn't get it done either. She did not offer to help.

I drove over to the sheriff's office and visited with the famous old sheriff, Buck Echols, who raised Longhorns himself. I told him what my problem was and how bad I wanted to interview her for the article. Buck, to say the least, was a little rough around the edges. He had carried a gun for a long time and was a no-nonsense, crusty old character. He scared me to death while I sat across from him at his large metal government desk. He insisted on calling Mrs. Russell and "cussing" her out in my behalf. While he was dialing her number, I begged him not to call. I did not know Sheriff Echols had been old friends and even went to school with Mrs. Russell. They had a crude way of discussing the issue—I could hear her shouting on the other end. He called her names involving ill-repute and nasty motives, using jailhouse vernacular with colorful adjectives that would heat a branding iron. He was destroying all hope with this lady that I had worked so hard to acquire. He hung up and said she was @#$%&! hopeless. That was that.

My next visit was with Wiley Knight, who had bought a Russell cow and bull in the last Raywood sale. He managed a grocery store a few miles north of the jail. I visited with him as people tried to check out their purchases while a boy sacked groceries on my left. Wiley also wanted to register his two cattle but was scared to try to deal with Mrs. Russell. She had her bluff in on him. We talked in hopeless despair for about 40 grocery-sacks time, then Wiley took a phone call. He said the call was for me: "It's Mrs. Russell. She wants to talk."

I will never know why. Maybe she needed a good loud cussing; maybe she thought I was never going to give up. Who knows? She said to come over after supper. We would look at old cattle photos and certificates and talk about the cattle. She finally acted agreeable. Something had changed.

As the sun was setting in Liberty, I drove south of town on Highway 563 to where Pauline Russell and her husband, Wiley, lived. The home had a big front yard with some flowers and tall pine trees. I took some photos of my cattle to show them. I was skeptically invited inside, where we talked in a sort of friendly cool way. I had spent a few hours on the phone with Pauline, so we had some familiarity with each other. The couple was in their 50s. Wiley was quiet, and Pauline talked about how mean Henry Butler was with horses and cattle. She was bitter about a number of things which had nothing to do with me. She had located some old registration certificates, mostly of cattle she no longer owned, and remembered some of the old bulls' names. It had been about 10 years since she and Milby Butler had attended Longhorn sales together in Fort Worth and Wichita Refuge.

I asked questions about Bevo. She said he had larger horns than the white bull that Blackie Graves had bought. That would make Bevo over 60". I took extensive notes about the Butler cattle, especially the interesting things about certain ones that she remembered. She talked of a great love of the cattle and of bad people who are mean to animals, and I agreed. She disliked Blackie with a passion. Each mention of him was viciously bitter. She never said why. I didn't ask. Of course, I knew.

The Beauty family was the cream of the Butler cattle. Beauty was Pauline's favorite, and she indicated Milby Butler wanted her to always keep Beauty.

We talked on until near midnight. I was very complimentary of the cattle she had raised and sold at Raywood. She acknowledged they were all her pets, but she had run out of grass and was sad they had to be sold. I assured Pauline that these cattle were the largest-horned cattle ever and deserved to become famous. My article would help them to be well-known by stating that she had raised them right in her backyard. But there was a very sad thing: none of the people who had bought her really good cattle had registered them. Those bulls would never have any calves registered nor would the cows. They were unregistered, grade cattle. I wanted her to know an unjust thing was happening and she was the victim, but I was there to help. Pauline listened, then said words I shall never forget: "That is wrong! They should register them! That makes me mad! Why don't they do it?"

I carefully promised Pauline that I would do everything in my power to MAKE them register all her cattle with her name properly on the certificates. It was important to me and the industry that all the really good cattle be registered. I just happened to have brought a pad of registration applications in the car, and I would help her fill them out. She was very pleased that I was ready to hammer these people to rightfully register her cattle. She determined that they had somehow cheated her again on these registrations. I kept my promise to "make" them register the cattle and to record the fact she was the owner and breeder of these great animals. I especially had to promise to FORCE Blackie Graves to register his two white bulls. I promised, offering to even pay the fee myself if he refused. He never paid. And I did.

It was dark as I went out to the rental car and got the application forms. I dropped them all over the driveway. Pauline and Wiley worked at remembering which calf was born first and when. They recalled returning from a trip when one calf was born. Beauty had another right after some family member's birthday. One was born the week of Easter. One by one during the wee hours of April 7, 1978, we recorded the times of birth in chronological order. I filled out the basic info. Wiley H. Russell was the legal owner of Bevo, and Pauline was owner of record for Beauty. Both had to sign each application. No cattle had names yet, so I wrote a description at the top of the application. For Blackie's bull, it was "OLDER GRAVES BULL."

When all the papers were filled out, the Russells were very appreciative. Pauline made Wiley go out in the dark, crawl down into the wellhouse pit with a flashlight, and bring up an old dried skull from a favorite twisty-horned cow named Queeny. It was a gift for helping them with the papers.

It was past 1:00 AM when I left the Russells and stopped at a 24-hour convenience store, used a pay phone, dialed Blackie's number, and woke him up. I told him the paperwork was done and the white bull would be registered within 24 hours. He was slowly waking up and could not believe it. After many months, I had finally earned the half-interest in a soon-to-be-famous white bull.

I had several names saved for future special bulls. I asked Blackie what we should name this one. My choices were Impressive, King, Conquistador, Monarch, Overwhelmer, or Classic. Blackie chose Classic.

I had worked on this project for over a year, and time was important. There was a lot of disease in the Gulf Coast area, so health testing was something that could be critical to the project. Booster Stephenson pulled blood on Classic at first light. I phoned the Longhorn association and recorded the name and ID number, and he received his designated registration number. Classic was officially registered. I mailed in the hardcopy the same day. Booster did the Bangs blood test in his vet truck, and Classic was negative. He was loaded into a trailer and on his way to Elgin Breeding Service at Elgin, TX, for semen collection. By 4:00 that afternoon, Carl Rugg, the Elgin technician, informed me the first collection looked perfect. In 24 hours, Classic was registered and frozen semen was in the tank.

Classic was born in 1973. The other family members included Reveille, a full brother, born in 1974 and owned by Booster; Lady Butler, a full sister owned by

Classic ($1,000,000) was the most valuable Texas Longhorn bull in history. He was bred exclusively by Blackie Graves for natural service, and everyone else profited from him by artificial insemination. Reportedly Blackie purchased him at Raywood Auction for $317.

Wiley Knight; and a son of Lady Butler we named Monarch, owned by Blackie. Knight also owned a full brother named Butler Boy, born in 1976. My promise was complete to Pauline. All were registered, and my Butler story was published in the *Texas Longhorn Journal* in 1979.

A 1-page ad in color was placed in the *Texas Longhorn Journal* announcing the largest-horned bull in history available in frozen semen. Semen sold at $35 per straw, and Longhorn lovers were buying it 20–50 straws per order. Being partners with Blackie, we split the gross sales dollars. I did the sales, handled the semen sales and shipping, and paid for the one ad. In the first year, Classic earned over $60,000 in semen sales.

When Classic's calves hit the market, the business got even more exciting. We raised a chocolate-speckled bull named Rural Delivery and sold him in thirds for a total of $103,000. The fanciest young heifer in history was born by Classic. We named her Sweet 'n Low and sold her to Betty Lamb for $116,000. Many other notable progeny excited the Texas Longhorn industry in the early '80s, creating the greatest value boom the breed has ever known. With few exceptions, the high-selling cattle all traced back to Classic or Beauty.

The long-awaited application to register Classic

As a partner with Blackie, my business had been very good. I was artificially breeding and doing embryo transplants using Classic semen and creating up to 300 calves per year. Blackie wanted to keep Classic in Dayton and hauled him to certain events to show him off. I didn't push for natural service use of him because AI was working well. Semen sales were still very lucrative the second year.

The Red McCombs spring Fiesta sale is always well-attended and well-advertised. The food is good, and Red is always a gracious host. Many cattle have sold at this sale for very enviable prices. During the '80s, I always attended. As I walked up the hill to the sale area, Red and Blackie were talking in a pickup and motioned for me to join them. Red laid out a plan that was amazing. He had a plan to syndicate half of Classic's semen in shares and exclusively sell 20 shares for $50,000 each. No one else could purchase the semen except shareholders. Classic would be the first One-Million-Dollar Bull in the industry. Red was a brilliant businessman, and normally whatever he planned to do got done. As I leaned on the truck door, my first thought was, how do WE split up the $1,000,000? Blackie said that he had sold half of Classic to Red, which was all right with me; but then

Sweet 'n Low was the first Longhorn ever known to reach 48" horn tip-to-tip by her 24th month birthdate. A birthday party was celebrated with a corn cake and 2 candles.

The Rest of the Butler Story

I was ordered to shut off all sales of Classic semen. If I wanted to buy a share, I would be allowed to do that for $50,000.

My gentleman's-handshake deal had worn out. Blackie had originally agreed to give me half of Classic for services rendered, but then he sold another half to McCombs and kept a half. Somewhere beyond the handshake agreement my half went away, never to return. My heart dropped.

The legendary Classic Semen Syndicate was introduced during the O. Wesley Box Longhorn Sale at Elbert, CO. From start to finish, all 20 shares sold before sundown—$1,000,000.

Classic is the all-time top sire of money-earning Texas Longhorn cattle. His timing was right. Nearly 30 years later, as pedigrees are closely researched, it is difficult to find valuable Texas Longhorns who do not trace back to Classic. He and his queen mother, Beauty, are salted throughout the high-value pedigrees. Famous modern cattle would not exist without the blood of the Russell family of Butler lineage such as Cowboy Tuff Chex ($165,000), Top Caliber, 3S

Classic was displayed at the annual YO Ranch Texas Longhorn Sale. As the sign shows, I was no longer a 50% owner.

Danica ($380,000), Kinetic Motion of Stars, Casanovas Magnum, Shadow Jubilee ($120,000), Winning Honor, Tempter, Sweet 'n Low ($116,000), Jamakizm, Over Kill, Super Bowl, Outback Beauty ($150,000), Drag Iron, Hunts Command Respect, Clear Point, etc. Classic influenced the breed by appearing in registered TLBAA pedigrees 54,343 times.

Although correct conformation, rapid gainability, and pretty-colored cattle are always very valuable. Classic started the horn explosion. The money and big-money people followed. ▷-D

Clear Win is the leading sire of over-90" T2T sons. Although his banding with multiple Phillips breeding has added thickness, he traces twice to Butler genetics, Classic, Superior, Bold Ruler, and once to Rose Red, Ghost, and Maybelle. Clear Win is a careful mating of numerous virtues that explode into one wonderfully correct package.

> *"Have what they want; make them want it."*
> ~ Red McCombs, on successful marketing

Guthrie Buck and Walt Hellyer

The famous cutting horse Guthrie Buck was born in 1958 at Guthrie, TX, on the historic 6666 Ranch. He was a son of Hollywood Gold out of a Joe Tom by Joe Hancock dam, and owned by Walt Hellyer of Brantford, Ontario, Canada. He commissioned an oil portrait of Guthrie Buck.

Walt Hellyer was the first president of the Ontario Cutting Horse Association. He loved cutting horses but made his fortune exporting Canadian ginseng to China. Walt had some beautiful rolling grasslands in the breadbasket of southern Canada. His horse fences went up, over, and across in every direction. He had many miles of oak fence, with nearly every board randomly "horse chewed." The day I arrived, I took photos of Guthrie Buck from all angles for anatomy reference and drew a sketch for Walt's approval of the proposed oil painting.

I had met up with Walt at the *All American Quarter Horse Congress*, which is always in October. He had problems with vandalism during Halloween, so he exited early from the Congress to get back to protect the ranch from local beginner terrorists. I rode back with Walt out of Ohio north into Ontario. He was a brilliantly entertaining man. As he drove, we talked horses, ranching, and cattle. It was late and we had driven way into the night when Walt's trainer complained that his belt buckle was scratching his backbone and he badly wanted to stop somewhere and eat. I didn't know where we were; we were just driving in Canada somewhere.

We watched for another 50 miles, but at that hour, not a lot of places were open. Finally a normal-enough-looking log building bar/grill/combo appeared to be open. There were some seedy Paul Bunyan types hanging around and some well-dressed clients. We ordered something big to eat, because the Congress corny dogs had not been all that great. We waited so long that we were ready to eat napkins with mustard. Finally the waitress said that if I did not take off my Stetson hat, she would never serve us. That was a new one for me. I asked what was the problem?

She pointed to a snippy old wrinkled lady who was staring daggers at us about three tables away. The waitress said that my hat was offensive to one of the clients. We were hungry. I know I was really hungry. Walt discussed with the waitress that we, too, were clients, and we thought everyone should also be wearing western hats. (I personally was ready to take my hat off without a fight.) Walt took a second to think about it and then told the waitress, "We were also badly offended—the food was probably made out of s**t." We quickly got up and left. If you are ever there, don't tip that waitress.

Guthrie Buck—30"x36" oil on canvas

Walt drove for hours and never found any place to eat. By then, it wasn't long until breakfast, and the Canadian sun came up. A new day started in a new country. ▷-◘

> *"It is good for a man that he bear the yoke in his youth."*
> ～ **Lamentations 3:27**

Wallaby

During the *All American Quarter Horse Congress*, I attended the first dozen or so. For the first few, I stayed in downtown Columbus at the YMCA. It was cheap. A room was $4–5 per night. It was my plan to meet the cast members, make a living, take photos of great horses, and occasionally sell a portrait commission.

I had a hard time getting great photos at the Congress, with Ohio rain, cloudy weather, bad backgrounds with electric poles, old dark-colored historic buildings, and very few open areas. The light from morning to evening made backgrounds change from east to west. Shooting was not easy. The one thing good about the Congress was that the best horses were all there. Over the years, the really great stallions came to the Million Dollar Stallion Avenue. People came from around the globe to see the great ones all at one place.

Grier Beam, a very successful trucking guy from North Carolina, had bought Wallaby for a lot of money. Wallaby was by Croton Oil out of a Sugar Bars mare. He was being hauled by Kenny Bacchus and starting to stack up a wheelbarrow load of Championship trophies. Kenny had just begun working for Mr. Beam and had never had such a good job in his life. He was doing everything he could to please Mr. Beam. This was big to Kenny.

Wallaby was born in 1964, and this photoshoot must have been about 1968 or 1969. Mr. Beam wanted the best for Wallaby. He wanted to advertise him a lot and needed to get a great photo for a postcard.

Kenny and I waited for the right light, but things weren't working out. Finally we decided to take Wallaby off the Congress grounds, out the south gate, and photograph him on a lawn near a large government red-brick school building. Wallaby was a beautiful horse, but he had seen everything, been everywhere. It was nearly impossible to get him to put his ears up or show any poise. We did everything to get his ears up. I was discouraged. Kenny had 2 or 3 helpers doing all kinds of crazy contortions, but nothing was working.

  I don't like to photograph a horse with a chain under the chin or especially over its nose. I don't like any chain to show. I asked Kenny to switch the chain from the left side of Wallby's halter to the right so it would not show. As he unsnapped the halter lead, the big stallion bolted and sprinted full blast in an easterly direction toward a slum-project area. He was going AAA right through the children's playground where kids were screaming, running, and basically doing their best to outrun the horse. I laid my camera stuff down, and we all ran in the direction Wallaby was leading—much slower, of course. He ran about a half-mile, through some clotheslines, cars, and tricycles—right down the pavement and sidewalks. I will say it was exciting. If you think Kenny was excited, so were the folks who lived there in the projects. Finally, Wallaby slid to a stop in the corner of a cul-de-sac and just froze, out of breath. It was like he wanted someone to come and save him. He was lost in a world of screaming kids running in every direction.

Kenny talked kindly to him, slowly snapped the lead on, and we walked back to the school lawn. Amazingly, my camera equipment was all still there. In minutes, the kids were playing like nothing had happened. When we got to the photo-spot, Wallaby had every vein bulging. He was on the "muscle." We took shots from every angle and used this side view for the postcards. He was alert. He gave us abundant alert ear shots. This sleepy, been-everywhere, done-everything stallion came alive.

As we were finishing the shoot, shockingly, up walked Mr. Beam. As I remember him, he was about 5'9", maybe 60 years old, businessman-looking, and a very nice fellow. He loved this horse very much. As he stood there savoring Wallaby, we all knew exactly what had taken place. Mr. Beam had no clue about the near-disaster that had just happened. No one brought up the subject. He said, "Kenny, you are good. I have never seen the muscles and veins pop out on Wallaby like that. Whatever you did for the photos, to say the least, you are professional."

Kenny smiled!   ▷-ᗪ

"The trouble with doing something right the first time is that no one appreciates how difficult it was."

~ **Kit Pharo**

Deep Bob and Empty Pockets

Deep Bob was a 1960 son of Depth Charge out of Bobbie Leo by Leo. He was purchased by a fellow who lived southeast of Denver, CO, and a serious breeding program started to blossom. Johnny Dial, Piccirillo, Super Charge, Brigand, Tiny Charger, Dividend, Chudej's Black Gold, Johnny Bull, Hijo the Bull, and numerous Depth Charge progeny were becoming highly popular. At one time, Depth Charge, Top Deck, and Three Bars were controlling factors in the Thoroughbred introduction to Quarter Horse bloodlines. Depth Charge was one of the great speed stallions owned by the King Ranch in Kingsville, TX.

The owner of Deep Bob hired Joe (not his real name) as his trainer/manager. Joe was excited about his new high-paying, stud-manager position and commissioned me to paint a portrait of Deep Bob as a gift to his boss, the owner.

Deep Bob was a dark seal-brown with pretty roan white hairs. He was a very hard color to paint as I basically had to paint the white hairs one at a time. No fun. Really no fun.

When the painting was complete, Joe acted somewhat funny, in fact, real funny. He ducked and dodged paying me. Eventually he 'fessed up that he had been fired. He was really sick of the job, hated his old boss, and had no fond thoughts of Deep Bob, the stallion, either. But he wanted to make it right and pay me for the painting. We were friends and he would never cheat me. He handed me a check in full. It was exciting to finally wrap up the deal.

I did my part, but the check bounced. I called his bank, and they said he did have money going through the bank—but it didn't stay very long. They couldn't hold a check while waiting for funds. They weren't much help. Of course, Joe was their client and I was not.

Several months went by—Joe moved again and started a new training business with several clients' horses.

Now I was always respectful to the stall-cleaners and feeders that I met at any ranch. When I was taking photos, a willing ear-getter-upper was extremely helpful. I sometimes gave out postcards of famous horses and simple gifts to these men. I was getting paid to get the best professional photos. They were very important to get an ear to flick at the right time or a leg placement in the exact correct spot.

I talked to one of Joe's stall cleaners and found out he was on a cash basis with the feed company. When they brought a truckload of horse feed, they insisted on getting their check before it was unloaded. Joe's feed checks must be cashing. The amount of the normal feed check was just over what he owed me on the painting. Being short on funds, it was obvious that Joe was more concerned that his ongoing feed bill get paid before I did.

One day the phone rang; my informant said he had just handed a check to the feed company. I beat it to Joe's bank, about a two-hour drive, and got my bounced check to clear his bank within three hours.

Some of this business has not been easy. ▷-D

Deep Bob—oil on canvas 24"x30"

" ... *the way of the transgressors is hard.*"

~ **Proverbs 13:15**

Don Quixote's Mystery

Homozygous is a "high-sounding" scientific adjective used in articulate circles. It describes "an organism with identical pairs of genes with respect to any given pair of hereditary characteristics, and therefore breeding true for that characteristic." In simple backwoods vernacular, *homozygous* means "If your momma is really ugly and your pa is really ugly, you are sure enough going to be ugly, and probably your kids, too."

This genetic DNA study can be easily understood. When a black Angus is bred to a horned Hereford, the resulting offspring is always black and polled, with a white face. The white face is homozygous from Hereford and the black and polled factor is homozygous from the Angus. This pairing of genetic factors never changes; it is homozygous.

In the Texas Longhorn breed, few homozygous traits have ever been documented. The horn, the browse utilization, the spot-pattern colors, longevity, and calving ease, although wonderful traits, can all be diluted when mated with families who are not carriers of these qualities. Using words like *never*, *always*, and *absolute* when talking of Texas Longhorn matings can make one appear foolish in this realm of "nearly homozygous."

This chapter deals with the facts of history and consistent genetic results. Texas Longhorn is a breed with a wide range of traits, all very different from other breeds and honed by the fine-tuning of many centuries. This breed teaches many genetic lessons for the coachable.

The closest to generational homozygous in the Texas Longhorn breed is the black genetic of the old foundation Wichita Refuge-bred bull, Don Quixote Spear E 113. His black DNA is strong enough so that, if it is linebred, a homozygous strain may have been or could be developed. Over the years, as often happens, his name was shortened to Don Quixote.

Don Quixote

Thousands of foundation cattle have edged into unknown generations, yet a certain few individuals remain that are now proven to be breed-changers. The Wichita Refuge (WR) has raised nearly 12,000 Texas Longhorns. The WR family of genetics can be found in well over 100,000 cattle, but only one WR line has held strong, dominating among all other families. For a variety of reasons, certain cattle changed the force of pedigrees and took control of the generations. Such was the case of Don Quixote. Some may argue about who are the "top" black cattle, and some confuse black with real dark-brown. Still, most believe all great black cattle trace directly back in an unbroken line to Don Quixote at least one or more times.

Many cattle are called black but really are a dark seal-brown. If a critter has rust-colored inner ears, a mealy muzzle, or even a slight rusty or dun tone down the back line, this would not be a true black. A true black is black from the nose hairs to the tip of the tail.

True black Texas Longhorns "nearly always" have one or more true black parents. A true black has "nearly never" been recorded without a black parent, unless one parent is a silver grulla, which always has a black skin. In this case, a black can sire a grulla, the offspring be grulla, and then another black can be born in the next generation from that grulla. Although there is a close relationship between the grulla and black, getting a black from a grulla doesn't happen often.

Lazy J's $48,000 black steer Bluegrass illustrates the strength of Don Quixote's near-homozygous black factor. His black lineage goes back to Don Quixote. Bluegrass was sired by the bull J.L. McBride's Pride, a true black. His black lineage traces through his dam to her sire, Diego's Hot Shot, who was sired by Blackwood's Diego, who was sired by Crown C Don Diego. Continuing back the black line, Crown C Don Diego was sired by Don Quintana, who was sired by old Don Quixote. That means that Lazy J's Bluegrass black line is a 7-generation black line starting all the way back with Don Quixote. Amazing as the strength of the Don Quixote black is, many of his lineage produce 70% black or black-speckled progeny.

The first black-base cow to pass the 70" mark was Sarasam by Quill, tracing an unbroken, 4-generation line back to Don Quixote.

Trophy's First Lady, a true black, has an unbroken black line through her dam, granddam, and great-great-grandsire Circle K Donovan. Circle K Donovan's black line leads back to Don Quixote through Signal, out of Right Turn, by Impressive, by Don Quintana, by Don Quixote. That's a total of 8 generations from Trophy's First Lady back to Don Quixote. The beautiful Night Safari BL833

Flair Galore, age 2, is a linebred Don Quixote still holding the tiny-speck pattern of the Don.

traces a black line to her sire, BL Night Chex, out of Miss Quixote Special, out of Quixote Joan by Don Quixote.

One of the widest-horned black bulls is Over Kill. He is the sixth generation from Don Quixote, with no other black animals in his pedigree. His black unbroken line through his sire is Overhead, by Headliner, by Archer 92, who is out of Archer 381 by Don Quixote. Over 70% of Over Kill's calves are black-based. That is not homozygous but is getting very close.

By going back through the aging records of Wichita Refuge, wildlife biologist Jeremy Dixon located data that showed on September 21, 1961, a calf branded WR 1882 and a heifer calf WR 1878 were purchased by Ted Lewis of Maud, OK. The bull cost Lewis $115 and the heifer $130. The bull was registered as black with white on belly, and the heifer was a walnut dark-brown. The sire of WR 1882 was WR 1156, recorded as a black-blue roan.

TLBAA records indicate WR 1882 and WR 1878 were registered by Oklahoman Archie Beller, then transferred to Elvin Blevins' Spear E Ranch. Blevins registered a black bull calf as Don Quixote Spear E 113, named due to his Mexican appearance. Blevins consigned him to the famous TLBAA promotional Centennial Trail Drive Sale in Dodge City on July 3, 1966. He was purchased by Ralph Chain of Canton, OK.

In the early 1970s, Texas Longhorn cattle were not a hot item. For many years, an $1,100 bull led the statistics as the high seller. Texas Ranger JP, a flat-horned bull raised by Jack Phillips, was the largest-horned bull (48.75" T2T) and also had the largest body in the industry. He was used by Dickinson Cattle Company during the early '70s and was the first TLBAA AI certified bull. His semen sold for a whopping $5.20 per ampule.

As Don Quixote's calves began to attract commercial interest, DCC was looking for the next cross on Texas Ranger daughters. With the Ranger family's record horns and sizes, his calves were often a pale dun, some had extra leather under their necks, and in general they had a coarse, sloppy conformation which often included large sombrero-spot patterns. While visiting with Oklahoma Longhorn producer Kenneth Archer, Ken and I discussed Don Quixote, the bull being used by Chain Ranch.

Jet Black Chex is used at DCC to perpetuate the Don Quixote legacy. He sires over 60% black-base calves and is 9 generations from Don Quixote.

We went to see Don Quixote. The bull was exactly what I felt would work on Texas Ranger daughters. He had a very straight back, a very trim underline, and a small symmetrical head with ears so tiny a baby's hands could cover them. Don Quixote was as black as four-foot up a stove pipe, with tiny white specks on his hind quarters. He was strong in every area where Texas Ranger was weak. All genetic successful matings are about the next antidote correction.

World history documents that four-fifths of all cattle were bred for use as pulling oxen. In his book *The Longhorns*, J. Frank Dobie carefully records that all cattle originally arriving to North American shores were trained pulling oxen. The blend of these early cattle, mostly from Spain, included various types, colors, and breeds of Spanish bovine. In this case, Texas Ranger more closely resembled the oxen type, and Don Quixote the Mexican range cattle.

On July 11, 1972, Don Quixote was purchased from Chain Ranch by Dickinson Cattle Company, LLC (DCC), then headquartered in Colorado. DCC immediately planned to collect his semen; however, the collection company was scared of the big bull and refused to collect him. Terms were finally agreed upon that they would collect him only if he were halter-trained and could be handled safely with a halter. We proceeded to halter-break the 7-year-old bull, which was not an easy job. When the day came that Don Quixote was taken to the AI center, he made one collection jump and donated over 800 ampules. That was his only collection for nearly 10 years. Most of his famous progeny came from that collection (priced at $5.20 per ampule). He was TLBAA certified as AI bull #2.

DCC bred Don Quixote for several years, mostly to Texas Ranger daughters. He was then purchased by Texas oil man Cy Rickel, Jr., and next by the Wright Ranch at Robstown, TX. He completed his service life at the Safari B Ranch of Vici, OK.

During Don Quixote's life, there were only a few hundred Texas Longhorn producers and very few of those used AI. From small herds, a handful of calves were produced. The better-known ones were B Bar D Texas Don, Don Quintana, Choya, Quixote Cheetah, Don Abraham, Quixote Joan, Don Quinado, Oklahoma Quixote, and Quixote Caledonia.

Saddlehorn, owned by Brett and Darcy DeLapp is linebred 15 times with 2 direct black lines to Don Quixote 5–7 generations back. He has tested to be homozygous black. He measures upper-80" T2T.

Shadow Jubilee is one of the most striking Texas Longhorn cows in history. She is a blend of Classic, Senator, Texas Ranger, Susan 28, Don Quixote, Conquistador, Beauty, and New Sam.

At current Texas Longhorn shows, as many as a third of the entries are black or black-spotted. With few if any exceptions, they all trace back to Don Quixote. Some are as far back as 12 generations—and still black. Most are multiple times linebred to Don Quixote.

Some highly respected cows with the foundation lineage of Don Quixote are Outback Beauty ($150,000), Shadow Jubilee ($120,000), Starlight, 3S Danca ($380,000), RZ Temptress, Shutterbug, and Jester (linebred 7 times). Bulls that share a piece of his foundation DNA include Cowboy Tuff Chex ($165,000), Clear Point, Winning Honor, Tempter, Jamakizm (5 times), Rodeo Max ST, Drag Iron, Casanovas Magnum, Rio Grande, Hunts Command Respect, Circle K Donovan, BL Night Chex, Saddlehorn (15 times), and Jet Jockey (TLBAA World All-Age Halter Champion). Don Quixote influenced registered Texas Longhorn cattle by contributing his black DNA 14,200 times.

Only a few people are living who actually saw Don Quixote in person. Through his down-line progeny, his dominant mark of black, dark colors, small specks, and correct trim-type forever lives on. Consider this: the early colors of registered Texas Longhorns were mostly dun or a pale red—dominant pale red. Therefore, had Don Quixote not contributed his bold-black-slick-coated DNA, there is no doubt that very few, if any, true black Longhorns would exist today. Without "The Don," it would have been a camouflage world of very pale, dishwater-colored Texas Longhorns.

"The prudent heir takes careful inventory of his legacies and gives a faithful accounting to those whom he owes an obligation of trust."

~ **John F. Kennedy**

Tonto Bars Hank and Walter Spencer

Tonto Bars Hank (Hank)—born in 1958 by Tonto Bars Gill and out of Hanka. He was raised by C.G. Whitcomb of Sterling, CO. The Whitcombs took him to his early race wins, including the All American Futurity. Later, Walter Merrick became his trainer. Hank was hard to train and would play around and goof off while other horses were running their hearts out trying to beat him. He drove Walter crazy; however, he was so strong and fast he would win anyway. Hank had a speed index of 100 and enjoyed 16 wins while racing with the toughest competitors. His winnings reached $133,919. He had 31 halter points and was an AQHA Champion and World Racing Champion for three years.

After racing Hank, Walter Merrick showed him a few times. Merrick was a great racehorse man and judged big shows, but he wasn't a showman like Monte Horn. Mike Coen purchased Tonto Bars Hank, and Walter Spencer of Tulsa got control of his stud bookings and promotion.

Spencer was not new to the horse business. He was a stern, broad-shouldered fellow and sort of bowlegged—but not as bowed as Elmo Favor. He always took the little things of life as an all-out battle.

To enhance his stature, Spencer showed Hank at the biggest shows. He made sure the judges knew the 3-time World Champion would be at the show. Once, at the Fort Worth Fat Stock Show, Spencer had him entered in the senior stallion class. To distract from Hank's less-than-show-style head, he had a halter made with a huge red sheepskin nose piece. As all the pretty stallions quietly walked into the show ring, Spencer stayed back in the chute area with Hank. This class consisted of over 30 stallions, all badly wanting to win Fort Worth. The announcer asked for the number of the Spencer entry to enter the ring. Spencer did not come.

He announced again; Spencer did not come. Others in the chute area were urging Walter to get out there, now. There were people in the coliseum who wanted to see this famous stallion who had bested the fastest horses in the world. They were eager to see if he also had the class to compete with this bunch of the world's best show horses. The announcer called Spencer's number over and over. Finally, he boomed over the loudspeaker, "Entry, #--- Tonto Bars Hank, last call to enter your class!" That was the special introduction Spencer had been waiting to hear. He ran into the arena with Hank, head held high, and went straight to the judge and apologized to the high heavens for being late. The crowd's cheering mounted to a roar as they watched this master showman. Their enthusiasm left the judge with very little choice—Tonto Bars Hank was selected Grand Champion.

Walter Spencer was a detective on the Tulsa police force. He drove a Cadillac and ate big charred rare steaks. Once I was going to dinner with him and placed my hand on a towel in the seat of his car. He grabbed my wrist and promptly showed me a huge chrome .45 revolver under the towel. He explained that he had sent a number of murderers to prison and some were being released. He always packed heat and kept guns hidden within easy reach. He had had some death threats and didn't plan for any of those to be carried out!

Spencer once called and had me fly to Tulsa to do a photoshoot of Tonto Bars Hank after he was retired to the stud. Hank was big—huge. He had a long back, a short and slightly thick neck, and a slight roman nose. He was a lot like his sire who, while not very pretty, was massive and impressive. Photographing him from either the side or front would be difficult due to his lack of refined physical characteristics.

It was a horribly hot and sultry Oklahoma day. Spencer held Hank, and a fellow named Sterling used a sprayer to control the flies. He also helped by positioning Hank's feet and getting his ears up. I always requested that a team help with pictures: a handler, a foot-mover, and an ear-attractor. All others who merely wanted to watch were required to stand a long way off.

For this photoshoot, we were shorthanded. Hank was a playful sort, moving, stomping, and doing anything else he could to aggravate us. Finally, after an hour of minimal progress, Spencer cussed and griped and said he had had enough of me; he would never have me take photos for him again. He handed the lead to Sterling and stomped off mad to an office/bedroom-combo building where there was an air conditioner.

In spite of the playful stallion, flies, heat, and now minus one helper, Sterling had the stamina to keep working with me. We were both soaked with sweat but continued to work with Hank for 2 or 3 hours. Sterling was trying to hold him and I was moving his legs. Hank never gave away anything for free. Each leg was a wrestling battle. This photo is the result of our efforts. We made postcards for Spencer by the thousands.

When finished with this photoshoot, I was dripping wet and exhausted. I went to the office where I dreaded facing Spencer. He had taken a cold shower, and I found him standing stark naked and spread-eagled over a floor air-conditioner vent. Conversation was not fun. He put some clothes on and quietly took me to the airport.

Three years later, I was at Jim Ray's Quarter Horse Farm in Decatur, IL, to take photos of the famous AAA stallion Bar Money. Walter Spencer called to tell me he was showing a chestnut stallion with 4 white sidewalls that had won 32 Grand Championships and he couldn't get a good photo taken at any of the shows. He wanted me to fly to Tulsa to do a shoot. I reminded him of his threat that he would never have me take photos again. He was in a good mood, obviously cooled off, and said, "You totally misunderstood. I said I would never hold a horse for you again. I am not going to either—Sterling will." ▷-D

"Capitalism is what people do if you leave them alone."
— **Kenneth Minogue**

Hard-Water Christmas

As my dad, Frank Dickinson cruised up into his 80s, he insisted on keeping a herd of cows and some horses. Every morning and evening, he would go out to feed, run water, and in the cold Colorado winters, cut and fork out thick chunks of ice. Although he complained to his old buddies, all ten years younger than him, about the work he had to do and how cold the north blizzard winds were, he wouldn't sell his livestock.

As he got older, he fell on the ice with alarming frequency. He would limp around, gripe, and complain about his injuries. Mom warned that if he kept falling, he would break a hip and die. Then she would quickly sell all his livestock. Her continued threats led to a period when Dad quit falling, or else just did not talk about it. Then we noticed he had purchased a thick, insulated, "North Pole" coverall outfit. We think he continued to fall but had so much padding he just rolled around on the ground and got back up.

When it was -20°, the water tanks would freeze 4–7" overnight. Dad wanted the cattle to have warm, fresh water, so he would break the ice and fork it out every morning. Over time, he would accumulate huge piles of solid ice chunks. As winter wore on, the piles got larger.

Dad spent hours doing chores. The ice-removal was especially exhausting. I told him my men would fix a continuous-flow valve with a non-freeze overflow buried in the ground. Then the cattle would have fresh water all the time, and there would be no ice to fork out. He refused. I offered repeatedly. It eventually became apparent that Dad refused to die in a rocking chair. He demanded of himself a couple hours of serious exercise every day. If we took this job from him, he would not get his "wonderful" exercise. Being the crusty rancher he was, he considered it embarrassing to be flaunting around the house doing acrobatic exercises. ▷-D

On Christmas Day 2000, I drew this cartoon sketch for Dad. He liked my drawings. It is a medium of ballpoint pen and crayolas, framed for his Christmas gift. Sure enough, when he quit forking ice, he passed away at age 88.

"Remember that guy that gave up? Neither does anybody else."
~ **Stevie J Moss**

Dealing with J.L. Collier

The precarious J.L. Collier of Gustine, TX, was a frequent phone caller during the exciting early efforts at embryo transfer during the late '70s and '80s. He called with questions about the famous embryo-donor cows that Dickinson Cattle Company (DCC) had worked so hard to accumulate. His favorites included Rose Red, Measles, Maressa, Ranger's Measles, and Doherty 698. He did not plan to have a lot of cattle but always seemed to have the money to get what he wanted for his small herd.

J.L. and his wife, Mimi, showed up at DCC in Colorado during the summer of 1985. He wanted to see every donor, every well-known sire, and the up-and-coming young herd sires being test-bred.

That summer was a time of brilliant new young bulls, mostly developed from the embryo program. The Colliers saw Bail Jumper, Overwhelmer, Superior, "King," and Phenomenon. Bail Jumper and Overwhelmer were beautifully colored and had great conformation. Superior had close-up, cork-screw genetics and was the largest weight of any Butler bull. But his frame was coarse. "King" was the first over-65" T2T horned bull to weigh more than a ton. His size was almost earth-shattering.

The Colliers liked Phenomenon, who was spending his first breeding season in DCC's Wyatt pasture with 28 cows. He lacked a smooth, trim conformation and straight back, but he had more horn for his age than the other young bulls. Phenomenon was not for sale, because his sire, Superior, was soon to leave and the Doherty 698 embryo collections were over. DCC considered Phenomenon important to continue that unreplaceable lineage.

J.L. was persistent. He wanted Phenomenon but went back to Texas without him. After a few dozen phone calls, DCC offered Phenomenon to J.L. with one important condition: DCC would retain non-exclusive semen rights and have the right to collect semen from Phenomenon at a future time convenient to his

Phenomenon—age 2, one of the largest-horned bulls of 1985, was test bred at DCC and AI semen rights were reserved for DCC use.

regular breeding season. The deal was done. Phenomenon finished the breeding season at DCC and moved to his new home in Texas, never to leave the Lone Star State again.

After the Phenemenon calves were born, it was obvious his semen would be useful, so it was time to collect from him. He was unrelated to a lot of the well-known main-line bulls. I called J.L. to plan for a convenient collection time. Out of nowhere, he determined that since he now had total possession of the bull, he would not keep his word. J.L. broke the purchase agreeement, and DCC was never allowed to collect Phenemenon's semen.

Jumping forward to 2004—Super Bowl, a big red-pinto bull raised by DCC, was at Elgin Breeding Center in Elgin, TX, for semen collections. A half-interest had been sold to Red McCombs. An adjoining pen held a very wide-horned bull named Temptations The Ace, owned by J.L. Collier. At this time, the Collier bull was one of the first bulls to go over 80" T2T, and several related females in the Collier herd were also believed to be of record horn growth. This was possibly the strongest horn family at that time. Temptations The Ace was double-bred Superior and Doherty 698. It made sense for DCC to buy semen and add this exciting young bull to the herd genetics.

My plan was to call and work out a deal with J.L. My phone call started out very pleasantly. I told him I wanted to buy 50 straws of semen. He said it would cost $100 a straw—was I still interested? At that time, $100 per straw was the highest price of any Texas Longhorn semen. J.L. went on to tell me what a great bull Temptations The Ace was and what a great job he would do for the DCC herd. I agreed to send him a check for $5,000 for the 50 straws of semen that we wanted to purchase. This was good timing, since the breeding season in Ohio had already started. But it also meant we needed to keep the insemination process moving quickly.

At 6:00 the next morning J.L. called me. He was short and blunt, "You don't deserve this bull! He is too good for you! I will not sell you the semen. 'Bye."

After hours of planning for this linebred bull to provide an outcross on the DCC herd and thinking about the great horn that could result, I was devastated. I paced the yard and kicked dirt. Everyone knew J.L. was difficult. Many had tried to deal with him without success. No one else was offering him $5,000. My deal had vanished with just one short phone call.

Plan #2: I called Bill Burton of Burton/Stockton Ranches and asked if Bill would call J.L. and buy the semen for me. Bill indicated he did not get along with J.L. that well and J.L. probably would not talk to him on the phone. He thought his partner, John Stockton, was the better choice—everyone liked John Stockton.

Plan #3: I called John. He said he knew only "one person" who could get along with J.L. and was willing to call him. This "one person" did the deal with J.L. with a quick phone call. Within minutes, J.L. transferred the 50 straws to the "one person's" account at Elgin. The "one person" then transferred them to John's account, and then John transferred them to our account. We called Elgin with shipping instructions, and the following day 50 straws of semen departed Elgin to DCC in Ohio.

DCC sent a $5,000 check to John Stockton, who forwarded a personal check for $5,000 to the "one person," who sent his personal $5,000 check to J.L. Collier.

Tempter is one of the larger-horned, larger-base, and large-bodied bulls. His daughters are horn tip-to-tip leaders in the industry.

On July 8, 2004, the first straw of Temptations The Ace was implanted in Field of Pearls, and on April 4, 2005, a spotted bull calf was born, later named Tempter. Within days, other great results came from this valuable semen, including Tempting Roses, Tempted Friend, Tempt Me Not, Light A Shuck, The Ace of Roses, and Temptress.

Tempter is now 87" T2T, with progeny in the U.S.A., The Netherlands, New Zealand, and Australia. After breeding at DCC for six years, he was sold to Burton/Stockton Ranches of Cleveland, TX. Through AI, Tempter has sired some great progeny, with two daughters projected to soon reach the 100" mark.

Since the Phenomenon and Temptations The Ace negotiations, hundreds of even more valuable cattle have been bought and sold by DCC. Most of the purchases are simple, friendly, and uncomplicated. In fact, to prevent misunderstandings, DCC provides every cattle buyer with a 2-page written guarantee. It is simple to do deals—with most people. ▷-◁

> *"Unless you are willing to be as unreasonable and as brutal as your opponent, do not engage him in a conflict—because he will win."*
> ~ **WWII Veteran**

The Bob Shultz Bull

During the *National Western Stock Show*, which was always held in the most miserable cold month of the year, January, DCC purchased a booth to market cattle and ranch products. Our family had a booth for 28 years and leased an apartment in Denver to house employees who traded off booth-work schedules.

Robert V. Shultz, a Denver businessman, came up to the booth one day and started asking questions about Texas Longhorns. He wore a city suit and had no appearance of a ranch-life experience. I asked him what he did, and he said he had an insurance company. He insured things no one else wanted to insure, and then he "really socked it to them." He told me he had once worked with Evil Knievel through Lloyd's of London for certain accident policies, before the daredevil's Caesars Palace-fountain cycle jump. Bob was discouraged not to get the policies approved. I think we were, from the start, probably really good friends as he never made a move to "sock" any high-dollar insurance to me.

Before long, Bob showed up at the DCC ranch, east of Colorado Springs and was dead serious about buying some cattle. He had purchased a ranch up in the high elevation of the Rocky Mountain foothills west of Golden, CO, and was ready to start building a herd. He started out buying ten economical cows and from time to time bought more cows from DCC and others. Soon Bob ran out of grass and bought another ranch near Elbert, CO. Later his best ranch was the Prairie Canyon Ranch near Franktown, southeast of Denver. Over the years, he bought Zorro Ranger, Bold Ruler, and also the famous sire Colorado Cowboy as a calf. He raised the bull Bouncer, who was later owned by DCC and by World Champion wrestler André the Giant.

From the first day I met the businessman Bob, he evolved to "Cowboy Bob" or "Dakota Bob." His city clothes changed to a huge hat with a large feather, boots, and total Old West attire. Once he showed up proudly wearing a new vest he had

Dakota Bob Shultz, mounted, and judge Randy Witte, *Western Horseman* **publisher and** *Horn Stew* **proof checker.**

paid a lot of money for, yet it had a worn, used look—he had paid extra for that used look. I joked with him that the bottom looked like a dog had bitten, ripped, and torn his fine leather vest. To Bob, that was a compliment. He liked the old rugged "brush popper" cowboy look.

The Texas Longhorns and the ranch that followed made Bob a cowboy celebrity around Denver. He was a fun guy and enjoyed people and his cattle. He talked about Blue Corn horses with tongue-in-cheek humor and enjoyed wearing old worn-out spurs around town at the major social watering holes.

Until I read an article about Bob in the **Denver Post**, I don't recall him ever mentioning his military service. A lot of old WW II veterans just don't talk about the war. Bob had volunteered for U.S. Navy Service in 1943. He was assigned to the battleship USS Missouri as an intelligence radio operator, right in the middle of the South Pacific's bloodiest battles. On September 2, 1945, the Japanese surrender ceremonies were held on the USS Missouri. Bob said, "When we got up that morning, Tokyo Bay was totally fogged in. Japanese foreign minister Mamoru Shigemitsu boarded the ship deck to sign the surrender. General Douglas MacArthur signed on behalf of the Allied Powers. The signing was over in a few minutes. In that short time, the fog burned off and Tokyo Bay was full of ships as far as the eye could see. The fog cleared out after the signing just as if everything was all wonderful and beautiful again." According to the *Denver Post*, Robert Shultz was one of the few living witnesses on the 70th anniversary of the Japanese surrender.

One day, Bob said he wanted to look at some cattle and arrived with his Dodge truck and trailer. He parked out under some trees and was friendly as ever. I noticed something in his goose-neck trailer, and he said he was taking a yearling bull from the Prairie Canyon Ranch over to the other ranch. The young bull was just riding along to be dropped out on his way home.

We checked some pastures where Bob asked about certain cattle but did not indicate he wanted to buy anything, just stopped by. He was always welcome. It started to get dark, so we drove back to the ranch office. I asked about the bull in the trailer and Bob said not to worry about it. He was keeping him; he would go home with him. We went to the local Bronco restaurant and ate supper. It was pitch dark when we got back.

In 1979, ranches all over had huge losses at birth, pulling calves, doing cesareans, and suffering major financial setbacks. It was a time when the medium-sized U.S.A. cows were being bred up with some really big European bulls. The huge calves were killing cows at birth, especially the first-calf heifers—the beginners.

I could sell all the easy-calving Texas Longhorn bulls I could get hold of; so could Bob. I asked him what he would take for the young bull in the trailer. He joked about how good the bull was and in his funny way led me on, but indicated maybe he could live without him. Finally, the teasing, joking, and all were aside. Bob "weakened" and agreed that I could have him for $950 dollars—but once the bull was unloaded he was mine, and Bob wanted to be paid before he was unloaded. This was unusual, but Bob was an unusual person, always with a smile. In the dim light, I could see the silhouette of the bull walking in the trailer with his back and head showing. The trailer was solid on the sides about three-quarters of the way up, so this bull looked like a pretty big yearling.

Ranchers were coming to buy young bulls every few days, and I thought I could make a couple hundred dollars on him, so we made a deal. I had to write a check and hand it to Bob. This was all done in the dark. He backed his Dodge truck up to the corral and opened the gate. In a second, I knew why he had come to visit. He had loaded his trailer with square bales of hay, on end, that elevated the floor of the trailer. The bull was standing on bales of hay about 20" tall to make a dark visual appearance like he was 20" taller. This little bull dropped out on the ground and was only about 30" tall. He was a runt—a dogie. He probably weighed about 200 pounds soaking wet, or less.

Bob laughed and laughed. He said, "Deal is a deal." I fed that little bull and fed him, but he would not grow. Every other bull was sold that year, yet he was still in the pen for sale.

Bob laughed and laughed some more. He told that story to his city friends and got a lot of mileage out of it. Some friends are just that way. I did not think it was that funny. ▷-◁

"In the history of the world, no one has ever washed a rented car."
~ Lawrence Summers

Joe Queen, Rogers, and Murphy

Joe Queen—born 1952. Sire: Joe Reed II. Dam: Queenie. Joe Reed II also sired Leo, Joak, and several other greats of his day. Joe Queen ran AAA and was a great reining horse. Unfortunately, no one like the famous Oklahoma horseman Bud Warren ever found him to promote his genetics.

In the early days, Quarter Horses were nested mostly in Texas. The bull-dog type rodeo and ranch horses often evolved out of Texas. When big money started to flow in the Quarter Horse racing industry, the spotlight shifted to the West Coast and California. Los Alamitos Race Course, right down the road from Disneyland, was where really big things happened. The industry was led in the Rocky Mountain States by the Wiescamp horses, who were ranch, race, and fancy-conformation horses. The *All American Quarter Horse Congress* attracted the business to the Midwest, and Oklahoma pulled a lot of the race industry out of California. However, with all the migration in the '60s and '70s, fast Quarter Horses first found fame and money in California. It was also the home of early performance reining horses.

Joe Queen

Ridden in hackamore by Tony Diaz, Joe Queen was a great reining horse. He was a hot-blooded athlete as evidenced by his rippled muscles. Joe Queen was a big horse who always photographed well. He had far better conformation than his more famous half-brother Leo.

Although photographing horses took me to all corners of the industry, each year I spent mid-April through May in California shooting the great stallions. I first photographed Joe Queen at the Double R Bar Ranch, where Roy Rogers and Dale Evans had lived and shot many of their films. It was a tight little canyon with no open space for a clean background. Their address was 22832 W. Trigger St., just outside of Los Angeles, in Chatsworth, CA. Roy and Dale loved kids; but gut-wrenching things happened, and three of their children died while living at the Double R Bar. Possibly due to some horrible memories, they sold off parts

Joe Queen

of their very valuable land. Little by little, the 300 acres was whittled down to a few dozen when I was there.

Audie Murphy had filmed a Quarter Horse special for Disney, riding Joe Queen in a reining demonstration. He had owned the horse and was standing him at the Double R Bar. For a breeding farm, it did not have a lot of horse facilities; but some of the Double R Bar backgrounds were recognizable from old Roy Rogers films. I did not meet Murphy, but his horse-handling people assisted me as I took photos for the Joe Queen promotion. Audie was the most decorated U.S. World War II Army hero and a true cowboy as well as a Hollywood western movie star. In 1971, he died at an early age in a plane crash near Roanoke, Virginia.

"I'm an American actor. I work with my clothes on. I have to. Riding a horse can be pretty tough on your legs and elsewheres."
*~ **John Wayne***

The Shadow Saga

The Shadow saga starts with Caledonia, the first Texas Longhorn heifer registered by our family. She came in 1967 nursing on the first cow purchased from Graves Peeler of Atascosa County, South Texas. Caledonia was large, with a rich wine-brown color and a few pinto-white décor spots. In the early '70s, she was bred to Texas Ranger JP and produced Caledonia Ranger, who was bred to the magic cross of Don Quixote to raise Quixote Caledonia. From Quixote Caldonia came Nifty Fifty-Five, who was bred to a bull the ranch purchased in 1979 named Colorado Cowboy, who sired Bellaire, the dam of Good Friday, The Shadow's dam. That is the female lineage from 1967 to June 6, 1991, when The Shadow was born.

In 1978, Red McCombs came to DCC in Colorado and bought his first semi load of registered Texas Longhorn cows. He badly wanted Caledonia and her daughter Boldonia, by Bold Ruler. Those were the two cows he could not leave Colorado without owning. Being a car dealer, he traded the ranch a beautiful new, fully loaded LTD Ford car for those two cows. It's okay to let a good cow leave occasionally, as long as her best sons or daughters are retained in the herd. Linda and I flew to San Antonio, attended a Texas Longhorn event, and drove back to Colorado in our new car. That first sale to Red McCombs proved to us that the Texas Longhorn business was good—very good.

A little bull calf sired by Senator was born to Good Friday, summer of 1991. He was plain red with a few white spots that weren't very visible. I erroneously considered him a roping steer prospect. He was not noticeable in the herd nor did he have a name or brand.

Tim McCollum, an attorney from near Tollhouse, CA, decided to purchase a set of Texas Longhorns that summer. He selected Good Friday with her calf and a half-dozen other cattle. Off to California they went.

In 1996, nature had taken its course, and now the little red bull was the big herd sire holding court over his mother, his sisters, and all his nieces. He was

Good Friday pictured as a young cow with The Shadow as a calf. Many of his progeny are similar in color pattern to his dam. This photo was taken near remains of a homesteader's wilted dream.

not registered, nor had he ever had a bite of grain, just California wild oats. He had slicked off and turned nearly black, actually a very dark seal-brown. Tim called and wanted to sell him, stating that he was nearly a ton and sported 70" horns tip-to-tip. That was quite remarkable and certainly caught my interest. I asked Tim to measure the bull again "carefully" and send photos. When they came, I was excited. The bull was beautiful, slick, dark, with a huge hip and horns stretching into two counties. I had to own him as soon as possible, yet that was not going to be easy. Tim, having seen less than a couple dozen Longhorns in his life, was certain this bull was the best bull that ever had nursed a cow, and he was highly valuable.

Tim also wanted to squeeze the most money out of him that was possible. He wanted $10,000 for a half-interest, wanted me to collect semen and market it, and give him half the gross proceeds. I had big ideas for this bull, so the complications of a partner who was not going to help carry any water was a definite negative. With give-and-take phone call after phone call all winter, the haggling finally came to an end in spring of 1997. We convinced Tim that it was not good to have the bull sire all the calves from his sisters, daughters, mother, and whole related family. He needed an unrelated bull for his herd. It was the right thing to do.

Attorneys get paid well to negotiate, especially when it is totally for their own interests. My one advantage remained that I was the only kid on the block wanting this black bull. Our give-and-take went back and forth like a tug-of-war. Tim was totally bluffing and I was completely desperate. Neither wanted the other to know. I didn't know if the bull was gentle, would stay home, or any exact details, but I had seen garden-variety photos that were acutely enticing. I had so much respect for his sire, Senator, that I knew what could happen with the strength of lineage clear back to old Caledonia.

So a deal was finally done. We sent Tim a bull named Zarth, by Zhivago, and some unrelated cows to clean up his herd-incest mess. Tim retained a small semen-percentage interest in his black bull. We were still not exactly sure about his weight or measurements, but from all accounts, the black bull was sure-enough a looker.

As the McCollum deal was jelling, George Lucas was purchasing DCC breeding stock to develop a Texas Longhorn herd at his Skywalker Ranch near Nicasio at San Rafael, CA. Logistics worked well for us to deliver cattle to Lucas and pick up our new black bull from Tim on the western slope of the Sierra Nevada range. Our son Kirk and Todd Van Natta, who worked for the ranch, delivered

The Shadow in Ohio

Shadowizm is one of The Shadow's many sons who are siring quality progeny to carry on his genetics.

the Lucas cattle and then on February 22, 1997, loaded up the big black bull at Tollhouse. McCollum had no interstate health paper nor a squeeze chute to draw blood, so Kirk loaded up the bull and took him to Dr. Sallie Phillips' Sierra Veterinary Hospital in Auberry, CA. Todd and Kirk put a rope around his horns, and Dr. Phillips drew a specimen from the jugular. The bull never made a bad move during the procedure. He was getting all the legal tests for transport, and a health certificate was prepared while Kirk and Todd waited.

As the truck headed back for Colorado, the bull had a 36'x8' trailer all to himself. The shortest route out of California to our ranch in Colorado was over Donner Pass, which is brutal in winter, with an average annual snowfall of 415". Our truck was stopped by authorities and not allowed to try the pass without chains—not only on the truck, but also on the trailer, due to its length. Kirk and Todd turned around and drove back down the mountain to a little town whose main income was selling high-priced tire chains. They bought chains all around, never to be used again after the Donner crossing. Worse things have historically happened in February on Donner Pass, so everyone on our team really came out pretty good. No one ate anyone.

When a dark, unbranded, unnamed, huge-horned bull stepped off the trailer at DCC in Colorado, I was on hand the minute his nose appeared. I was thrilled. We were all thrilled. When you love and esteem great genetics, an awesome feeling comes from seeing the next increment of the breeding plan falling perfectly into place, a special feeling like, say, receiving an Academy Award, like Pat Garrett after his encounter with Billy the Kid, or like driving the golden spike at Ogden in 1869. True cattle people who respect and appreciate great cattle can come right to the edge of passionate insanity when they spot a truly majestic one.

When the unnamed bull placed his feet on our soil, I followed him around the pen. His ears were small like a big fox. He walked wide on his hocks, unlike most Longhorns. His back was straight as an arrow, his muscles rippled as he walked, and he had a big round hip like his sire. His face was perfectly symmetrical with no long, homely nose like some of the old foundation cattle. He had a wide spread between his eyes. It looked like he stood right at 63" tall and had a body as long as a presidential campaign. He looked around and slowly, confidently walked over to a concrete bunk and started to munch green Colorado alfalfa hay.

The Shadow was honored on the *TLJ* Herd Sire Edition cover for January 1999.

He was at home, for the rest of his life. He was smart. He didn't have a mean bone in his body. He was all and more than I had expected. In cattle or horse deals, that seldom happens.

He weighed 1,910 lbs., later up to 2,100 lbs. when mature. His horns measured 63½" tip-to-tip during his 6th year. In February of 1997, this was data almost unheard of in the Texas Longhorn world. We rushed him off immediately to Rocky Mountain Sire Service near Denver for semen collections.

Before collection, he had to be registered. After searching the dictionary, going over name

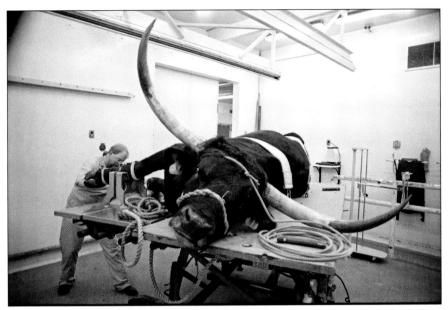

As The Shadow lies motionless on the Ohio State University surgical tilt table with his body belted like a straitjacket to prevent movement, Dr. David E. Anderson probes the hoof bifurcation for signs of permanent damage two months after the fiberglass toe was applied. Our good ranch vet, Dr. Harold Kemp, built the first fiberglass hoof.

notes, and quizzing others about a special name for a very special bull, Lana Hightower, an International Texas Longhorn Association (ITLA) judge, came up with "The Shadow" after a popular movie of years past. And "The Shadow" it was.

When the first collection of his semen was ready to ship, serious producers quickly placed orders and inseminated their cows. Before any of The Shadow's natural calves were born on our ranch, several dozen semen clients had his AI calves bouncing across their pastures.

In the Texas Longhorn business, a number of prominent bulls are white. As a result, over the years about 80% of the breed is dominantly white or over 50% white. Although countless light or pale colors and dominantly white cattle have multiplied in the breed, an antidote especially valuable to darken the base color and reduce the white was needed. That is the special genetic spot The Shadow was ready to fill. He was given over 80 mostly white cows for natural service in the summer of 1997.

In the registered cattle business, success is directly related to the quality of the herd sire or sires. If you do not have genetics right on the cutting edge of the highest quality, everyone knows that you are not trying hard enough. That makes a public statement, loud and clear. It is wonderful to own the best bull in the breed, but equally good is using semen from the top bulls. These top-end cattle offer large sale values, the middle-range cattle are profitable, and the bottom third are hamburger. All are a business, but some are far more profitable than others. It is important to work the right genetic direction. When The Shadow calves started to run the pastures, they were obviously the front-cutting-edge kind.

The Shadow sired dark and spotted calves. Due to his brindle ancestors, we got a higher percent of brindles than we had expected. We bred mostly white cows to him, but not all were dominated by his darkening ability. The first year, both fancy females and males were born. Very few bulls are strong enough to breed great progeny of both sexes. That is an arduous proving ground.

The Shadow was highly respected in the industry and offered a special look, style, and a much-appreciated gentle kindness not often found in herd sires. Semen sales were the best of any AI sire. Many serious producers wanted a son of The Shadow for their own herds.

He had almost completed his second full breeding season in September of 1998 when it happened. In a matter of only a few days, his weight dropped 200 lbs. and he spent most of his time standing in cool mud or water. When he came out of

Marshall Dickinson on a trained riding steer, Sniper, a son of The Shadow

the water, his right front leg was badly swollen, and he was in severe pain as he walked. We called the vet, a highly respected large-animal practitioner Dr. Harold Kemp, who told us that The Shadow's prognosis did not look good. The tissue on his foot had become necrotic, and the sole had started to separate to a serious degree. An infection had entered the bottom of his foot and was going up his leg. He had stepped on something sharp that penetrated his hoof from sole up to his ankle. Amputation or hamburger were the first two unpleasant recommendations. A third, long-shot choice was to remove the sole of the punctured toe, rebuild it with fiberglass, keep him pumped with antibiotics to kill the infection, and hope for the best. Dr. Kemp immediately proceeded with the long shot. If our big bull healed properly, his hoof would redevelop in about nine months—but only if he survived the infection already advancing inside his leg.

The Shadow was a good patient. We medicated him daily, enforced minimal activity, and designed a spacious stall for his very personal care. He walked placing his weight on the left front leg and carried his right. Most of the time he lay down. He became so used to shots and medications that Joel could go into his pen, lift up his foot, and treat him without constraint. He was a perfect gentleman. Little by little he began to put weight on the damaged hoof. By breeding season he was rested, well, and good to go.

When The Shadow was nearly healed, Paul Harvey included his brief story and near-miss of amputation in a "Rest of the Story" radio broadcast. For years, fan mail came to The Shadow

Bogus Bonus, a son of The Shadow, stands by his sire's memorial stone as in tribute. The Bible verse on the stone is Psalm 112:9c: "… his horn[s] shall be exalted with honor."

The Shadow Saga

wishing him a speedy recovery. So many people knew about him, the post office delivered mail to the ranch addressed only "The Shadow, Barnesville, Ohio." He became the bull everyone wanted to see when visiting the ranch. He weighed just under 2,100 lbs. and now measured 65" tip-to-tip. His dark, slick hair coat was striking as summer guests watched him survey his herd. When he walked, the ripple of sharply defined muscles was his admirable trademark, the sign of a true athlete.

The Shadow's calves soon began to enter prominent herds around the nation. DCC shipped his semen to New Zealand and Australia, where producers glady received it.

John Thate of Minnesota purchased a group of cows from DCC. In the group was a cow named Pueblo (purchase price $1,400), who was bred to The Shadow. Her heifer calf, Pueblo Shadow, was consigned to the McCombs sale as a 3-year-old and topped the sale at $17,000. Later, Joe Valentine bought her in the Legacy sale for $16,000. He raised 3 brindle heifers and sold her to Mountain Creek Longhorns for $48,000.

Ranch tours for the public are available during the summer. The Shadow was a favorite for guests. He personally ate "cow candy" from the hands of an estimated over 9,000 friends.

Mark Hubbell purchased The Shadow's semen and raised an awesome cow named Hubbell's Shadow Kay. She sold to Bill and Judy Merideth in the Legacy sale for $57,000. Numerous progeny of The Shadow sold privately for impressive, often undisclosed, prices.

Progeny of The Shadow increased in major herds all over the nation. Shadow-izm, a son, has earned a highly respected spot among the great sires of the breed. Mile Marker, another son, is a leader in siring horn-show champions. Shadow Rula is a near-duplicate of his sire. Other great sons are breed leaders: Bogus Bonus, Shadow's Reflection, Sombrah, Black Shadow, The Most Wanted, Texas Tornado, Guardian, and several more. Shadow Pomp was the leading sire of ITLA Élite and Honor Roll progeny. Shadow Savvy was probably The Shadow's largest son, weighing 2,080 lbs. when he turned 3, never being fed any grain. Shadow

L to R: James and Barbara Steffler, Lapeer, MI, with Shadow Jubilee's original owner, Joel Dickinson, Barnesville, OH, representing DCC. Steffler's cow Shadow Jubilee won the highest award in the Texas Longhorn business, the TLMA Ultimate Longhorn Cow. In the 2010 World Championship show at Durant, OK, she measured the widest horn spread of any bull or cow in the breed at 86⅛" tip-to-tip.

Spear won the TLBAA Horn Showcase with 96" tip-to-tip, a record for a 5-year-old steer at that time.

As The Shadow continued to do his job, tragedy struck again. It was a dry summer in Ohio, and the gravel roads near his pasture were limestone. As the local coal trucks pounded the stone 24 hours a day, dust drifted all over the pastures. Some pastures were gray with dust-covered grasses. As The Shadow grazed, he infected both eyes with limestone dust and went completely blind. It was the middle of breeding season, and his service was urgently demanded. Every four days we drove him with Honda 4x4's to a corral about a half-mile away and gave him antibiotics. Although he stumbled and tripped, slowly he made his way where directed, then returned to the cow herd. Animals are astute and seem to understand distressing situations. The cows apparently recognized The Shadow's disability and made every effort to cooperate. The following year's crop of calves was nearly 100%. Apparently good vision is not necessary to be a frequent father.

Months after the lime-eye fracas, The Shadow's eyes remained a hazy blue-gray. He was apparently blind. About six months later, I tried to get a photo of him. The men were driving a herd of cattle about a half-mile away up on a hill—just dots on the horizon—when The Shadow quickly turned his head, gazed intently at the moving cow herd, and directed his hazy eyes to them. His sight was back to some degree. I don't know if he could count the cows, but he could see them and he had a great interest. That was a joy to behold.

The Shadow's left front leg had carried most of his weight, protecting the injured right front—but at a price. He started to get arthritis in his left knee. As he worked to do his job, walking up and down the Appalachian hillsides became harder each year. We gave him some phenylbutazone pills in a daily grain mix to minimize his pain and provided every kind of care. In their line of duty, the hind feet of a bull take a beating with rocks and stumps. Finally, his left hind toe broke. He hobbled around with one good remaining hind leg.

As we watched, The Shadow faded away, showing increasing pain and difficulty standing. Our hearts hurt for him more every day. It was not a happy time. This famous old sire had the best healthcare that we could provide, but he departed this life on April 24, 2003. The Shadow was dedicated, yet totally used up.

We prepared a large memorial stone from the wall of a 220-year-old stagecoach stop building on the ranch. Then we placed it on the lawn near the DCC headquarters with an etched black silhouette of The Shadow, his date of birth (6-6-91), and no departing date. Through many miracles of reproduction, after his death his calves will continue to be born for years. The Shadow had the most progeny registered in ITLA and still affects the breed with 7,467 documented TLBAA herd-book entries.

Soon after his passing, a tribute was printed in the NCBA publication *National Cattleman*. He is one of the few bulls in history with full-color obituaries, and again Paul Harvey acknowledged The Shadow's death on his national broadcast. No other bull has received an obituary with a listening audience of 16 million.

June 7, 2003, The Shadow's memorial service was held at Dickinson Cattle Company north of Barnesville, OH. The Shadow was one of the most famous Texas Longhorn sires in history and a favorite of thousands of summer tour guests. Officiating was Dr. William Mummert of Gettysburg, PA. A host of The Shadow's friends gave personal eulogies to the greatness of his life and service to humanity. Cattle producers from 11 states and Costa Rica attended the service.

The Shadow Saga

> **Funeral for a Friend, by Paul Harvey**
>
> On June 7, a memorial service was held at Dickinson Cattle Co., Inc., in Barnesville, OH, for a famous black Texas Longhorn bull known as "The Shadow." Cattle breeders from 11 states and Costa Rica attended the service conducted by Dr. William Mummert of Gettysburg, PA. It was a private service, with eulogies presented by close friends of The Shadow near his memorial stone that references Psalm 112:9c "... his horn[s] shall be exalted with honor."
>
> The Shadow was born June 6, 1991, a 64-pound dark-red bull calf. Over time, his red color shed to reveal a very dark-brown color that looked black from a distance. He spent his carefree youth in California eating wild oats and eventually weighed more than a ton and sported a total horn measurement of 82".
>
> As he matured, his huge body, great massive horn spread, and wonderfully correct conformation became well known in the Longhorn industry. His first group of calves arrived in the spring of 1998; and while everyone was optimistic, no one realized how well-formed and valuable his offspring would become.
>
> At the time of his death, many regarded The Shadow as the leading Texas Longhorn sire in the nation. It is believed more of his semen has been distributed than of any other Longhorn bull of the last decade. His semen sold for $25 per unit. At $25 per one-half cc straw, his semen, per ounce, is valued three times higher than gold.
>
> The Shadow is survived by more than 800 progeny in the U.S., Canada, Brazil, Mexico, New Zealand, and Australia. Semen will continue to be available from The Shadow. He will live forever in pedigrees through his progeny, who will continue to be born annually as a result of the miracle of frozen semen.

One day before The Shadow's ninth birthday, a pretty red-speckled daughter was born, later registered as Shadow Jubilee. Like her sire, her red hair shed off to appear black, as pretty as a speckled pup. She grew in grace, frame, and horn, often to be selected as a choice purchase for buyers at DCC. All offers were refused.

Shadow Jubilee is one of the most striking Texas Longhorn cows in history. She is a blend of Classic, Senator, Texas Ranger, Susan 28, Don Quixote, Conquistador, Beauty, and New Sam.

Bill Burton and John Stockton first projected that Shadow Jubilee would be the youngest cow to cross the magic 80" tip-to-tip horn mark. They selected her for the new scientific cloning process that can create identical DNA duplicates. Burton, Stockton, and DCC struck a deal and extracted a tiny DNA specimen from her to develop clones. They then contracted ViaGen genetics lab in Austin, TX, to produce a total of 15 clones. Each would be a heifer and genetically the same as Shadow Jubilee, with slight speck-pattern variations. The clones were

These are clones produced from Shadow Jubilee. Each has her exact DNA and is registered as sired by The Shadow with the same pedigree as Shadow Jubilee. Although the cloning process was—and is still—very expensive, it allowed DCC to sell Shadow Jubilee and still have her exact DNA in clones for future generations.

The Shadow Saga

born between 2006 and 2008. Today, DCC owns most of the Shadow Jubilee family, while Burton/Stockton Ranches owns 4 clones and their progeny. The clones are all past 80" and some already surpass 90" tip-to-tip like their clone dam. Soon, The Shadow will quite possibly have more 90" progeny than any sire in history.

In May 2010, Shadow Jubilee was sold for $120,000 to James Steffler of Michigan. She left a family of over 20 members at DCC, mostly black-speckled, with more on the way. Steffler, known for his energy and positive appreciation of the breed, offered heifer-embryo pregnancies pre-sold for $20,000.

DCC has produced fine cattle full time for over 50 years and seldom seen a bull like The Shadow. But when one like him comes along, really good things can happen. These special "sports" have the ability to sire well beyond their own phenotype, going up to unusual degrees of quality. And thanks to the modern miracle of frozen semen, more Shadow genetics will come in future years, for DCC and for those who choose to perpetuate the lineage of this phenomenal bull—The Shadow. ▷-D

"You have to bear in mind that Gene Autry's favorite horse was named Champion. He ain't ever had one named Runner Up."
~ **Gene Mauch**

Moolah Bux

The Thoroughbred stallion Moolah Bux was a 1952 son of Mahmoud, out of an own daughter of Man o' War. He was tall and had classic conformation with very straight, correct bone and leg structure. Like his famous grandsire, he looked like he could run forever and never be unsound.

When the first Moolah Bux runners hit the Quarter Horse tracks, I was commissioned to go to southern California and take a series of photos of him at a famous Thoroughbred farm. He was tall, flea-bitten gray, and very gentle.

When I arrived, I found I had only one handler for the photoshoot. The fellow was very nice and really loved Moohah Bux, but unfortunately he was heavy on the sauce. He was sitting under the shed row watching for me and insisted I join him for a drink. I declined the offered snort, which seemed to severely hurt his feelings.

It appeared my shoot might be a wreck. I had a wonderful horse to photo, but my only helper was drunk. Ideally, I would have been assisted by an ear-person, a holder, and a foot-mover. This nice old fellow was so drunk it was sad. To his credit, he had patience and kept staggering around for hours, "helping" me in spite of his disability. Fortunately, Moolah Bux was a likable and

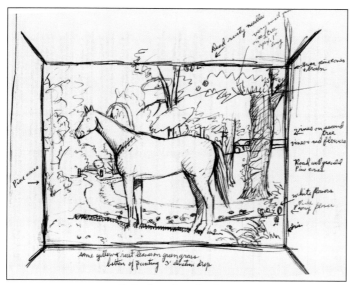
Painting sketch plan

Moola Bux painting—oil on canvas 30"x36"

gentle horse. He just wanted to eat grass rather than pose. The photos came out okay, but that great stallion deserved better.

Years later, I did a large oil painting of Moolah Bux for a different owner in Indiana. He never paid me. I don't know if the guy died, went bankrupt, or was just a crook. I do know he refused to give the painting back when an AQHA judge and friend, Bud Alderson, went over, knocked on his door, and tried to repossess the painting for me. Bud was sort of on the small side, so that didn't help.

If you know where this painting is, I'd sure like to get it back! ▷-D

> "People who think laughter is the best medicine apparently have never had morphine."
>
> ~ **Cathy Hawke**

Rhonda K and Silkay—The Process

Tommy Manion, one of the world's most serious equine competitors, often had me fly to Illinois to photograph his show string. For years, he leased the stalls and arena at the Illinois State Fair grounds for a few dozen up-and-coming prospects. His clients' horses were taken to the infield and posed for photos. It often took hours and sometimes days. When the photography was all done, we engaged in horse talk. We discussed the phenomenon of "finding" some nuggets on the race track and bringing them into the Quarter Horse halter arena. Some show people were finding these already-successful horses. The more versatile their competitive records, the more value they developed. That made the next step of their career development potentially very rewarding.

In the early '70s, a number of AAA runners were coming off the track. Some stallions became AAA, AQHA Champions and bred a lot of mares. They quickly took the spotlight away from heavy-muscled, thick-bred halter horses. The Three Bars look and the Custus Rastus look started to be very eye-pleasing to the judges. Dad (Frank) and I had been breeding Silky Fox, Skip's Count, Sage Scooter, Monty Carlo (full brother to Skipper W), and later Sonny's Calvin. We decided we needed to migrate toward the more Thoroughbred athletic look with more hot blood. I decided to find a half-dozen chestnut mares for Silky Fox and Skip's Count that were speed-looking horses. I bought a AAA mare named Sherry By Bar and Queen o' Clubs J, who was the dam of Royal Bar. We already had one Three Bars daughter and over a dozen Shoemaker mares—mostly sired by Sailor Cue.

The next goal was to find a couple extreme Thoroughbred mares with classic conformation. I went through the shed rows at the New Mexico State Fair, the Colorado State Fair, Stewart's Thoroughbred Ranch at Rye, CO, and Corky Keen's Keenland Acres in Fort Collins, CO. I wanted to find mares with four white stockings and perfect balance. Their race record didn't concern me; I was searching for mares with physical characteristics such as small refined heads, a long flat hip,

Rhonda K was extreme in length of flat hip, shoulder slope, and balance. She was chestnut with four white feet and all the good Thoroughbred eye appeal.

as much gaskin as you can get with a Thoroughbred, correct legs, and general correct conformation to blend with our stallions. Although I looked at several thousand horses, the ugly heads hanging over the stall doors eliminated most. It didn't take but a few seconds to glance into a stall and see every kind of foul anatomy and unaligned wheels. Most of the process was just walking the miles and miles of shed rows. It was as challenging as hunting for an honest politician. I didn't want bays or browns. Grays would have been good.

I bought a Thoroughbred mare from Les Davis at the CS Ranch. Her name was Carbise, and she was running claiming races at Albuquerque. She was a granddaughter of Battlefield, a bright-red chestnut with four whites and the most gaskin I had ever seen on any Thoroughbred mare. She was streamlined, and I thought she was the special mare to raise a great one by Skip's Count. In spite of trying, she didn't.

Another Thoroughbred mare, Rhonda K was a dark-chestnut born in 1958. She was the dam of one stakes winner and had a long-legged stud colt at side. I first saw her at Corky's Keenland Acres. She had the excessive frame I was looking for—long neck and flat hip. She was an older broodmare, but you could see the extreme athletic structure all over her. Corky wouldn't let me have Rhonda K until the foal was weaned, then he kept the colt. I took Skip's Count the 3-hour drive to Fort Collins. We bred her on the 9th day, and she had a long-legged palomino filly the next year. I would have bought more like her if I could have found them.

Rhonda K was bred to Silky Fox in the spring of 1973, under what I will call a "non-consensual" happening. She was so important, but her right "timing" for breeding was not easily determined, and we just could not catch the right time. A young veterinarian with a mobile lab palpated Rhonda K, but his test was non-conclusive. He recommended giving her a shot to force her into heat so that she could be bred. He was filling his syringe, certain it was the thing to do. I

Silkay coming age 2, just before she started to be unbeatable at shows. Everything was there—it was not all popping out yet.

Rhonda K and Silkay—The Process

Silkay was slick as a kindergarten doorknob, tall, 4 white sidewalls, and never burned out—always alert and looking around during a show.

had a bad feeling and told him not to give the shot. Later, we had mixed feelings about withholding the shot, since we never caught Rhonda K at the right time over the next few weeks. Finally, during another vet check, she was found to be pregnant from the "non-consensual" event. That offspring was to be Silkay.

In 1974, Silkay was born. Dad looked at her for hours. She started out really tall and gangly. She was chestnut, with the desired four white stockings and a blaze. Dad couldn't wait to show her, but she was too slender to place well. Although she was very tall and flashy, fillies with more muscle always beat her when she was a yearling. Some people who had been riding Silky Fox progeny tried to buy her, but Dad wouldn't budge. He had an offer a month—good offers to purchase Silkay—but he refused to sell.

As a 2-year-old, with a lot of Dad's brushing, Silkay blossomed. With the Thoroughbred blood, her maturity just kept getting extreme, and it held strong. Gary Campbell, who had ridden Skip's Count for Dad, took Silkay to the *All American Quarter Horse Congress* and won Grand Champion in 1976. Dad was especially proud when she was shown by Margaret Orr and won Grand Champion at the *National Western Stock Show*. He sent her on the road with Tommy Manion, and she was almost unbeatable.

Most of the Quarter Horse world knows of Silkay, AQHA World Champion Aged Mare. She won 133 AQHA points for both halter and performance, and a pickup-load of trophy toys. The really good ones like her do not come easy. That was the process of creating the beautiful Silkay. She made my dad and all those who showed her, including Ralph and Rene Levin of Sturgis, MI, who purchased her from Dad, very happy. ▷-D

Dad showed Silkay at a lot of Colorado and Kansas shows. She was a great joy for him. He took it seriously. Gary Lake photo

"... a sword is sharpened, and also polished; it is sharpened to make a great slaughter; it is polished that it may glitter."
                                        ~ **Ezekiel 21:9–10**

Hauling With or Without Class

In the book *Fillet of Horn,* there is a story of Red McCombs coming to Dickinson Cattle Company (DCC) in 1978 and buying a semi-load of cattle. The family car was so old and worn-out that Red caught his pants on a projecting spring in the passenger-side seat and ripped a big triangular hole in his hip-pocket area. That old station wagon had shown cattle for many pasture miles and had its problems. What a major embarrassment!

In the registered livestock business, there are no rules as to how to show pasture stock or what kind of vehicle works best. How to show cattle is not part of a college ag-business degree. As I have traveled Canada and the U.S.A. looking at cattle, I have watched a lot of marketing techniques, all designed to sell cattle to prospective buyers in some degree of comfort, more or less.

Years ago, most everyone would take buyers around the pasture in a pickup—sometimes very clean and other times with log chains on the floor, old mail on the dash, and ropes, a rifle, and a cattle prod hanging from the back window. Some upgraded to a dually truck with 4 doors and more room, but some duallies would bounce your dentures out in a rough pasture. Once Bob Coffee took Ben Gravett, Paul Babington, some others, and me in the back of a pickup on his really rocky Georgetown, TX, pasture. We were bouncing in the pickup bed working to keep aboard.

Big operations like the King Ranch use Suburbans. Lasater Ranch uses a Suburban, and Virginia Beef uses a Jeep Wagoneer. Penn's Cave uses a rugged Hummer. Some have pastures that require 4-wheel drive at times. It is not good to get stuck on the back side of the ranch with a legitimate buyer.

Three times I have bought cattle horseback. Two times it was pouring rain, and we were in up to a foot of water. The other time we were high on a mountain out of Grand Junction, CO, with a lot of downed timber and no roads. That was the only way to get to the cattle.

Clients buying quality cattle enjoyed not having to ride the pastures in a dirty old pickup. They were comfortable and treated like important people should be treated. This is client communication that can't be done in the auction-sale process.

Once I was shown cattle by helicopter south of Kingsville, TX, with Frank Horlock. Earl McConnell showed me a lot of cattle from his small plane over eastern Oregon. John Brittingham of New Mexico provided his customers seats in the back of his pickup—old car seats. A pickup will hold 3 of those, maybe 4.

DCC has not always been well-blessed with fancy vehicles. As Texas Longhorns became more profitable, instead of spending hard-earned money on fancy cars, we normally bought more land and cattle. For a couple years (many clients recall this), our temporary plan was to go to the airport to pick up prospective cattle buyers in a rusty old pickup, leave it in the rental-car lot, and lease a brand-new Lincoln Town Car for the day. Back then, if you brought it back full of gas, it was $45 per day. We could show cattle all day, keep clients comfortable, and not have them straddle ropes and tool boxes. We treated them in a professional way that they truly deserved. Sometimes air-conditioning was nice. We found a Lincoln Town Car had a great suspension and could handle rocks, sage brush, and rough terrain. Everyone was comfortable. When we picked one up, it did not have old food or kid's toy remains—it was nice and clean-smelling. For a personal touch, we slapped a magnetic Dickinson Cattle Company sign on each side.

After looking at cattle for sale, we took buyers back to the airport, retrieved the rusty old truck, and drove it back to the ranch. When there was one promising visitor a week, the 1-day rentals worked fine. It was fun to have a different-colored

Lincoln every day. Later, as business increased, there might be 3 or 4 buyers per week, or even more. At that point, it was more practical to lease a Town Car by the month. Later, purchasing with a long-term payout was even better, as we used this car a lot.

People enjoy feeling like they are "high-rollers." Some may be more apt to buy when treated with extreme comfort and elegance than in just an old ranch truck bouncing them around. If they are not comfortable, they will most likely cut the buying trip short—especially people in the party that did not want to be there anyway.

In the early '80s, we formed the Holt-Dickinson Ranches in Winfield, Alberta. My partner, Bret Holt in Canada, did not want to be low-class, so he showed the partnership herd in a Mercedes for his pasture car—if the snow wasn't too deep.

Back then, DCC needed to increase national awareness. Texas Longhorns were a minor breed with real upside potential, but many ranchers did not realize the genetic value and calving ease of the breed. We needed to move forward on national and international promotion. Peggy McDonald was editor of **BEEF Magazine**. She had won numerous awards for her writing skills, marketing, and photo journalism. **BEEF** wanted to keep her, but we were able to move her to Colorado with a nice salary increase. Peggy created ranch press releases and articles that were used by major publications all over North America.

We needed to catch up for lost time, and Peggy thought we had missed a major opportunity by not being involved in the National Cattleman's Association (NCA). She designed a trade-show display booth for the Phoenix, AZ, annual NCA Convention. There were 600 other display booths, so how could we stand out when competing with multi-million-dollar companies and their zillion-dollar displays?

In order to move fast, Linda and I, Charlene, Peggy, and Patty Jones flew to Phoenix 2 days before the convention. We leased 4 Lincoln Town Cars and slapped magnetic signs on the doors. We parked in front of Phoenix Sky Harbor International Airport where arriving convention participants would be looking for a cab. When anyone under a Stetson walked out the door, we offered them a free ride to the hotel, if they were attending the cattle event.

As many as five passengers per car, with luggage, headed to the hotels. The cattle people would ask who we were and would guess—the Phoenix Chamber of Commerce, the Welcome Committee, etc. We told them we were Dickinson Cattle Company and explained that we knew they would be busy during the convention, but we wanted to talk to them for 20 minutes about Texas Longhorn cattle. We would give them a free ride just to listen to our sales pitch. (A cab would have cost them $20–30.) Each Lincoln hauled up to 50 people per day for 2½ long days. As a result, over 400 cattle people knew exactly who we were and had received our promotional info. Most came by the sales booth during the next four days. We gained several new clients from the Phoenix Airport caper.

After renting one car per client, we finally bought a pasture car. Charlene Rogers Semkin was an energetic sales gal. She liked to fly through the sage brush with clients. Left, Jack Montgomery was a wonderful friend. He bought Measles and a number of embryos over several years.

The car rental, fuel, and all costs of this promo effort was about $3–4 per passenger. The sales girls did a great job hauling and selling, and, in fact, some passengers committed to bull purchases even before reaching the hotel. (Don't try this without some caution. By morning of the third day, the local cab drivers' union was steaming mad and threatening to ram us off the roads! It was too late.)

Ohio is a different deal. When the ranch relocated to Belmont County near Barnesville, the creeks, rocks, hills, and mud are not friendly to a Town Car. It rains a lot in Ohio, so visitors may need the protection of a covered 4-wheel drive—a Suburban. On nice days, which is most of the time during May through

From April through October, thousands have learned about ranching and the cattle industry on DCC pasture tours. Those of us in agri-business must tell our story to those who use our products. Otherwise, someone will tell them to just eat spinach.

October, a 4-seater 4x4 Ranger Crew UTV puts clients right nose-to-nose to view the cattle. It is up close so people can hand-feed if they want to. Not classy, not real comfortable—but visually perfect and great for photos.

Things change. In Ohio hundreds of people want to look at the cattle and also hand-feed them, but only a few are interested in buying. That is why DCC's ranch tourism started. For just $12 per person, a group receives the 75-minute narrated tour in a colorful, bouncy ranch bus. It is a rugged ride they will never forget. Some have returned 6–10 times to see different pastures every month. Guests can view hundreds of cattle each time, often different ones in different pastures in different months. Thousands of visitors actually survive and enjoy it every year. A percentage will buy cattle. A larger number will purchase high Omega-3 Texas Longhorn freezer beef at the Longhorns Head To Tail Ranch Store. Things change. Every target is a moving target.

> "Robin Hood didn't 'steal from the rich and give to the poor.' He stole from a corrupt government and gave it back to the overtaxed citizenry."
>
> ~ **Emily Zanotti**

The Partlow Purchases

Hunting cattle in the swamps, cactus, and brush of south Texas engulfed my mind and energy during the 1970s. I was sure the world's greatest bull or cow was in the next pasture, if only I could find it. Sometimes the cattle were very cheap and other times as high as a cat's back. Some of the owners were so wealthy they wouldn't even talk; others would fabricate any yarn, no matter how outlandish, to make a sale. Every deal was a new experience, with unique people and often with pleasant kindness, hospitality, and good food involved. Texas people know how to eat good and feed well—if they like you.

The Sam Partlow experience was very cold at first, one pizza-on-special-at-99¢, but that was the first of the week. Had I not stayed so long trying to buy cattle that Sam didn't want to sell, maybe he would have appreciated my company more.

Going back—In an old Texas Longhorn sale catalog, there was an ad with some heavy-horned white steers in a yoke. It was an R.G. and Sam Partlow ad. I wrote Sam and asked if he had any Texas Longhorns for sale. About six months later, he wrote a postcard that said, "I don't have any Longhorns and none of them are for sale."

I knew there were Butler cattle mingling around the pastures east of Houston from the liquidation of the Milby Butler herd. I had heard Partlows and Butlers were friends. I was sure cattle were out there in the trees, and the natives knew where they were. Blackie Graves, who had Classic, knew every waterhole in the county, every rice field, and every cow trader who ever spit in a bucket at Raywood Livestock Sale. Yet, he wouldn't tell. One by one, Blackie would buy swamp cows for $250 at Raywood—why bother to tell anyone where they were?

Pauline Russell had relaxed around me and was more friendly than many of the other natives. One day she called and said, "I found you some Butler cows. They have really big horns." She didn't know who the owner was, but she told me exactly the pasture they were in. I packed some clothes and headed to the airport.

In a few hours, I was driving a rental car east from the Houston Intercontinental Airport, a road that I had traveled often for the same reason. Sure enough, I found the cows north and west of Liberty on the south side of a farm-to-market road. Pauline said she thought Sam Partlow owned several thousand acres in that area and the cattle might be his.

Sam lived in a large old southern home in Liberty, TX. In his yard were tall pine trees, fig bushes, moss-draped live oak trees, and every kind of blossoming plant. He had an old Cadillac in the drive by his house, so I parked nearby. When Sam came to the door and I told him who I was and my interest, he admitted to having a few cows, maybe 20. I reminded him of the card he had sent and questioned his message. He said, flatly, he didn't want to mess with me. Right away I had a feeling we were getting off to a rocky start. I asked him if his Uncle Dode Partlow had any Texas Longhorns and would he sell them? He said he had suffered a stroke and was bedfast. The last thing he wanted to worry about would be doing cattle deals.

Sam had enjoyed a good life, was a little chubby, had seen some serious action in the military, showed no signs of western or cowboy attire, and was partly balding with white hair. At some time during his life, a lot of oil and timber money had gone through his fingers. He was not friendly, had nothing to do, and didn't want to do it with me.

In order not to waste his time, he invited me to ride around town to run errands. He showed me the laundromat, the big live oak trees in the area, the sheriff's office, and all the important things in town. He wanted me to realize what a great little town Liberty was. I had been there many times before, but I worked hard to act impressed. He pointed out several species of early-blooming flowers, and they were nice, considering I had come from Colorado where it was blowing snow when I got on the plane. We drove around, and he showed me a small pasture with some small-horned, hungry cows he said were owned by Sheriff Buck Echols. The sheriff had not repaired any fence since the election-before-last. If his cattle got out, a deputy in a black-and-white car would eventually put them back in.

Sam pulled up in front of a feed store, bought a sack of cottonseed hulls, and loaded it into the back seat of his Cadillac. We talked and drove, and in a few miles Sam got out and unlocked a gate with oil company No Trespassing signs all over it. We went down a gravel road in his pasture with standing water a few inches deep in little swampy puddles on both sides of the road. Sam got out and started calling the cows. He was looking toward some large trees, honking his horn, and calling, but nothing happened. We drove another quarter mile, and he called some more. A big red Brahman bull came walking toward the car, and slowly a few very hungry cows followed. One by one, up to about 20 cows from 3–15 years old slowly came out of the brush. Sam shook the sack, but they just looked at him. Being curious, I asked what he gave them for supplemental feed in the winter, and he said, "You just saw it." He poured little piles on the ground and cautiously some cows came up and ate. Mostly they just stood in the hulls and mashed them into the mud. Sam's cows were not used to any kindness.

Sam Partlow was a difficult man to deal with. His years of breeding a special line of Butler cattle became the foundation of many modern high-value cattle.

One pretty red-speckled cow with corkscrew horns was a spitting image of Beauty. Another was a walnut-brown with beautiful corkscrews. A few were pale duns and solid reds. They were really poor. I wondered how they could live until spring grass. One cow was a dark-walnut color with a big white splash in her face. She looked younger than the rest and was about as large tip-to-tip as the older cows. This was the largest-horned set of cows I had ever seen, similar in type to other Butler cows, but with more horn and less white. I wanted them. All of them.

Sam started his reverse sales pitch. He said Texas Longhorns were something that had been in his family forever, and he would always have some of these cattle. He said they were worthless and you couldn't get a decent price for them. No

This young, dark-walnut-colored cow, although hungry, became the foundation lineage for thousands of cattle with excellent investment values. This is how Maressa looked the first day I saw her.

one liked the horns. But he had learned that you could breed them to a big Red Brahman bull and the calves would bring up to $250 at Raywood. He said no one liked the cows but him, and if he sold them they would be doing good to bring $300 each at the sale. I asked him what he would take for the whole herd, but he wouldn't give a price. I was trying to hide my excitement, because I thought right here and now I am about to buy 20 pregnant cows for $400 each and we would both be happy. Then I started to think about those little flop-eared Brahman calves to be born in the spring. It would be a wasted year for me and the cows with their mongrel calves.

When we talked about a purchase, he refused to price them, for days. He would say, "What do you offer?"

I have always believed when you have a cow and someone wants to buy her, you have a right to place a price on your own stock. It can be high if you don't want to sell, low for a quick sale, or whatever. When an owner won't price his inventory, he may not know what they are worth. He may be afraid he will price them too low and leave money on the table. He may worry about being too high and scare the buyer away, killing a deal. Normally, when a person won't price cattle, it is difficult to do a deal. If I make a low offer, it may offend him. I may make an offer well above what he would have taken and pay too much. It is so much easier if the seller will just give his price and go from there. Sam wasn't going to be that easy.

I had just sold some cattle including the famous cow Measles and had a little over $20,000, in the bank. This was a chance to get some Partlow cows with more horn than Measles, and the closer to $400 apiece the better. I asked if he would take $400 each, and he was as offended as if I had tossed mud at his face. This was crazy. He had just said they would not bring $300 at Raywood. But, unfortunately, we were not at Raywood Livestock Auction.

After some cool but hospitable chat, I departed to a motel defeated and discouraged. The only satisfying part of the day was an old seafood place on the road to the Raywood sale barn.

The next morning, I went back to Sam's big house. It seemed like I was in his way. He was about 65 years old, a bachelor, but had relatives all over the area. There was a lot of money in the area and Partlows were a big part of it. Sam had a crazy aunt that he checked in on. We took her some food and talked about flowers blooming and birds building nests.

Then we went down the road a half-block and met Uncle Dode Partlow. I had read about him in an early Longhorn book. A warm, pleasant man with a lot of mileage, he was in bed and could hardly speak. His hands and face showed a long life of hard farm work. We talked and he seemed to be knowledgeable, but he couldn't speak clear enough that I could understand him. We talked about cattle, and his face brightened up. There was a well-used old leather KJV Bible by his bed that obviously some family member must read to him. I knew I would never see him again, but would have loved to talk cattle with him before his stroke. I asked Sam if it would be all right to pray for Uncle Dode, and Sam was sure he would be pleased. I prayed for his well-being and asked the Lord for any kindness that could be extended. Uncle Dode held my hand and shook with a

The Partlow Purchases

feeble grip during the whole prayer. He was a good old man. He was a cut above, and everyone who knew him knew it. That was a solid generation.

Then we drove south of Liberty to some rotten old barns, swamps, trees, and Uncle Dode's hungry cattle. There was a wooden gate that anyone who could lift 200 pounds could easily open, dragging it through deep mud. Several inches of rain had flooded everywhere. Bogged in Trinity River delta mud were about 150 cattle of all ages and sexes. There stood the tri-color brown-speckled Butler bull with high horns I had seen in Blackie Graves' pasture. He was the one that had been beating up on Classic. Uncle Dode had owned him all the time, and he had been used by Graves, on loan. The Partlow calves being born were sired by this bull. The cows giving birth were mostly sired by this same bull. Everyone was related—way too much. But these cows were different from Sam's: they were pretty, dark-spotted, higher-horned, and larger. Someone had been feeding them something, because normal swamp water and tree bark is not a balanced diet.

As we drove away from Uncle Dode's pasture, I asked Sam again what would he take for his own cows? He asked, "What would you give?" It was apparent he did not have a number in mind and had no clue where to start, yet he continued to endure me. We drove back to his house, and I said I needed to get to the airport and go home. I thought he would decide it was time to fish or cut bait and offer a sale price—wrong. First, he didn't need the money. Second, perhaps he was looking forward to those Red Brahman calves in a few weeks. Who could know?

So I gave up. I said, "If I could scrape up the money, would you take $1,000 each for those cows?" He said he wouldn't even consider it. "What would you take?" And he said, "What more can you offer?" It wasn't my first uncontrollable stampede, but he was making this herd hard to catch. I had a certain amount of money, and I needed to make it stretch as far as possible.

Finally, we worked up to a $2,000 offer apiece. It took about an hour that day, plus the previous four days to get this far. In this area, no Longhorn cattle in history had ever brought near half that amount. We seemed to be there, then he said, "No, you can't have them all. Which ones do you want?" We went back out to the feed store, got one more sack of hulls, and drove back to the swamp. As I selected the best cattle in the herd for my $2,000 offer, everyone was okay until

I picked the walnut-colored cow with the white-splashy face. He said, "Not her. She's the best I've ever raised. She stays here." I offered to take a total of 8 for $16,000, but I wanted her in the deal. Sam's answer was No! We worked her up $500 per notch, and finally at $4,000, he said, "Okay, she's yours."

I could see him backing out and ending the deal, so on the hood of the Cadillac, we wrote out a bill of sale. It wasn't safe enough for a handshake. I gave him a check for $3,000 down, contingent on every cow testing negative for Bangs and Sam furnishing the proper paper work for all to be registered. We had arrived. I would provide a check for the balance when the trucker from Colorado came to load. All of a sudden we were friends … really good friends. Just like that!

We went into the big house. Every step on the ancient hardwood floors made a creaking sound. There were high ceiling fans, elegant old stuffed chairs, things all around that 75 years ago had cost big bucks. Sam dug out dusty old boxes of records, photos, receipts, canceled checks, you name it. We hunted through old papers. Amazing as it was, a cow named Rose Red was properly registered. Miss Liberty 49 had papers, and little by little he found the old bulls that had sired the different cows. Some were full sisters. Certain cows you could tell were closely related with strong similar conformation. Sam knew when each was born, at least to within a year or so.

As I enjoyed the old photos, Sam had an idea. He said, "Now that we have this deal done, do you want some more cattle?" I admitted the truck coming down could hold more, so I was interested. He said he had a niece or some distant relative who had been given a calf by Milby Butler years ago, one that now had huge horns. He felt sure I would like her. The owner, Laura Scarborough, and her husband were building a new house, so it might be a weak moment for her. She might let this really good cow go if Sam talked to her just right. Laura worked at the local Production Credit Bank and he would call her there.

Sam told me to go to the front of the house and sit in the love seat by the front door. I did. He went to the very back of the house where he had a phone on the wall by the back screen door. It was a big house and had dark old walnut-color walls with a hall down the middle. I couldn't see Sam, but unknown to him, I could hear every word.

The Partlow Purchases

The call went like this: "Laura—Sam Partlow. You won't believe what I got going for you. There's a guy from Colorado that just bought the highest-priced cattle ever in Liberty County. He's loaded and dumber than a stump. He ain't got a brain in his head. He wants to see that cow Milby gave you, and I pumped him up to $2,000. He's ignorant enough to buy her, and I know you need the money for your new house." I couldn't hear the other end of the call.

Sam listened for a while, hung up, and came down the hall with the report. He said Laura wanted more for the cow, but he talked her down to $2,000. She was at work and it was not easy to get away in the afternoon, but he had also talked her into leaving work. She was putting all the details together, and Sam would take me to her pasture.

Laura lived north of Liberty and was saddling two horses when we arrived. Her pasture was flooded with water about a foot deep and she had no corrals. The only way to see the cow was to ride through the water horseback into the back pasture.

Maressa really brightened up with worming and normal nutrition. She produced No Double, the sire of Unlimited. She had several embryo calves by Impressive, Texas Ranger, and Classic that appear in many valuable pedigrees.

I got to ride a pony about 13 hands tall, and the saddle was one her kids used to barrel race. I more than filled it up, to say the least. We rode east into water from a half-foot to a foot-and-a-half deep with my stirrups and boots in water most of the time. Laura rode ahead into the trees, and a herd of mixed-breed Brahmans came stampeding out, splattering water so high they almost didn't see me. One cow running by had several brands, was a pretty grullo with some pinto-white, and appeared about 10 years old. Her horns were huge, as big or larger than Sam's cattle. Her registered name was Majestic, and she was that. Laura wanted the full $2,000, and I had her sign a bill of sale guaranteeing her to be free of Bangs. They had tried to catch her before, unsuccessfully, but Laura promised that if she couldn't catch the cow, I would get a full refund.

While I was riding in high water, before driving back to Liberty, Sam had been on the phone. He said that if I needed more cattle, Uncle Dode wasn't long for this world and his son Richard "Dick" Partlow would sell the Dode herd. Interestingly, when we got to Dode's, Dick was leaning on his pickup wearing tall mud boots, ready to show cattle.

Dick was part-recipient of an inheritance that Uncle Dode was dispensing. He had been feeding and caring for the cattle, but these daily chores were wearing thin. He had no special love for the breed and could use the money. He was a tall fellow, dark hair, pleasant, and motivated to sell—right now. He put out sweet mule feed in some wooden troughs half-buried in mud. The big, strong cattle quickly came to eat, while the poor, young, and smaller ones watched from a safe distance.

Dick and I sat on a feed bunk while the cows milled around hoping for more sweet feed. This was one of the first pastures I had seen with just one bull—the bull Mrs. Russell had told me about. The story was that Milby had blackleg hit his herd just before his death. Several cattle died. The whole herd was vaccinated, but this bull got away. Everyone thought he would die, but he did not. The bull was unusual with the prettiest blends from chocolate to near-black, with white specks, rust, some grullo, and gold. Many of the cows were colored like the bull, plus some were black Dalmatian and various real flashy colors. Most did not have the wide lateral horn of the Partlow cattle. They were high and twisty-horned.

An old cow named Miss Liberty 10 was demanding senior rights around the feed and looked nearly old enough to vote, but amazingly had a fat calf at side. A black-spotted cow walked across some deep water to a little island area about an arena length away. She appeared to be in labor with a "telling" crook in her tail.

Dick was ready to sell at $2,000 for adults if I would take them all. I didn't know what Sam had told him, but I could guess the script. In the next hour, I talked about what they would bring at Raywood, yet Dick couldn't get out of his head whatever Sam had told him. I inventoried the cows on a tablet. Some were pairs, some were ready to calve, and some were only dreaming of a future family. Some were crippled and well past voting age. There were steers mixed in, young bulls I didn't need, and of course some Brahman steers. I made a much lower offer than Dick wanted, by groups, and figured what could be hauled to Colorado on a possum belly. With cows, calves, yearlings, and the big bull, we were around 65 head. I had cut out inferior stuff that looked like flukers. I gave Dick a check for $2,000 as earnest money, did a bill of sale, and promised him the full balance with the trucker after the cattle passed their Bangs blood tests. That would take about 10 days.

Before we left the pasture, that pretty black-spotted cow had given birth to a black-and-white spotted bull calf. He was falling and trying to stand up in a few inches of water. His mother had lain down and given birth in a few minutes and never made a sound. She was licking him off clean and loving him.

When Sam could not tempt me into more bargains, I left for Colorado. Had he recalled any more relatives with Butler cattle, he would have called them.

The really hard part of the trip was telling Linda that I had purchased $63,000 of cattle, all poor, wormy, and hungry, with $20,000 in the bank. My borrowing ability at the banks was never easy for buying cattle. Bankers, all being on the fearful side, found my requests easy to decline. I had about a week to dig up over $40,000 before Ellis Gaddy, the trucker, left. I sent checks in the exact amounts for Ellis to hand over to Sam and Dick before loading.

More than just my rental car was in the area. Unknown to me, other people were quietly hunting big-horned Butler cattle, too. They wanted these genetics and were driving the same roads. Some had real deep pockets. Most of them went to Blackie, who would say he didn't know where any cattle were; he had not seen any in the whole area.

All my purchased cattle were gathered in some sturdy corrals for the good large-animal vet Booster Stevenson to tail-bleed them for shipment. Red McCombs of San Antonio, who was really getting into the Longhorn investment and breeding business, got a tip about cows on a farm-market road northwest of Liberty. He wasted no time and found the Partlow cattle being worked in a corral under some huge live oak trees. He drove into the pasture, got out, and leaned over the fence in awe of the horn he was seeing. It was an accumulation of the largest-horned cattle he or anyone else had seen. Red motioned for someone who looked in charge (I never heard who he talked to) and asked, "I would like to buy these cattle. Who owns them?" The answer was "Some guy in Colorado."

Years ago, a check took 3–6 days to clear a bank. There was a few days float time due to processing checks by postal mail. If you lived several states away, cash funds took longer to clear, but I still had not scraped up the money to cover my

Sweet 'n Low at age 2. She was considered the élite of the breed. She was a Champion show winner and led the nation in horn tip-to-tip measurement—48" at 24 months. She was by Classic and out of Rose Red. Sweet 'n Low sold to Betty Lamb for $116,000 in 1984.

checks. My plan was for Ellis to get the cattle on Saturday, when the banks were closed. That would not allow time for the deposits to hit until Monday morning. I had been telling people about the cattle, and several wanted to come see them when they arrived. I knew they would look lean and poor coming off the truck, and it would be better to have them viewed after they filled up on some good high-elevation alfalfa hay. I needed to sell enough to make the checks clear my bank. So I had no choice but to sell quickly the first two days after they arrived.

The semi was loaded clear full. Many of the cattle were small and lightweight, so a lot of them were on those two decks. The trip was about a thousand miles, but the truck got to the ranch before I had returned from checking a windmill. When I arrived, the driveway was full of trailers and pickups. The cows were eating good hay in our main corral. Every critter had walked off the trailer poor, but alive. There was an excitement over these Butler cows, and everyone wanted some.

The beautiful cow Shenandoah, who had calved while we were doing the Dode herd deal, had her calf fat and going. He was big and caught the eye of Bob Shultz. Bob purchased several cattle including that pair and named the little bull Colorado Cowboy, who later appeared twice in The Shadow's pedigree and also sired G-Man. Stan Searle bought some of the best ones, including the corkscrew-horned Miss Liberty 49. She became a good embryo donor for Stan. Jean Wickland and George Ansley each purchased 4–8 cattle to add to their herds. We loaded them into trailers and those sales made the checks all clear. The Butler family of cattle provided new genetics to add more horn to a breed already famous for its long horns, and hungry for more.

The first embryo flushes from the Partlow cattle were a heart-breaking disaster. We believed that an experimental treatment of repeated injections of vitamins A, E, and B12 would increase their vigor and fertility. It took many months, however, for them to get healthy, clean out their flukes, and get them up to normal reproductive flesh.

For embryo transfer, DCC used Maressa, the walnut-colored, splashy-faced daughter of Miss Liberty 49. Maressa produced No Double, the sire of Unlimited, who is believed by many to be the top sire in the Butler line. Rose Red proved to be a wonderful mating with Classic and produced Sweet 'n Low ($116,000) and Dixie Hunter. Many believe Dixie Hunter was the best siring son of Classic. The same mating was so profitable that 22 embryo calves were created. Orders were placed for future Rose Red pregnancies at a price of $12,000 each. At one time, 28 of her projected pregnancies were sold with deposits, but she did not produce long enough to fill all the orders. These same orders were in such high regard that some who ordered them sold their un-produced embryos to other investors. If an order was 18th in line for transplant, serious business types would pay a premium and purchase one in line at a lower number.

Miss Liberty 10 was 20 years old. She was sired by Liberty Lad, a Graves Peeler bull used by the Partlows. Several of the Dode cows were from her and her daughters. Dodes Boy is out of Miss Liberty 10. She had a life production of over 21 years, and even produced 4 embryo calves after she was old enough to legally drink. Quite possibly she had more fertile longevity than any cow of any breed.

The multi-colored bull was registered as Conquistador. He had sired calves for Uncle Dode for nine years, and for two seasons for Blackie Graves. Unfortunately, "Real Sam" received credit for his calves raised at Blackie's. Semen was collected on Conquistador, who provided his legendary genetics to Butler family appreciators. He was bred one year at DCC and sold to Haythorn Land and Cattle Company in Nebraska.

Today the Butler family of Texas Longhorns is one of the largest of the 7 original families. Due to court cases, family fights, poor management, and lack of concern, these wonderful cattle were very close to riding out of the history books. Although the Butler cattle were all highly linebred, some people love them enough to continue breeding them in a totally closed pedigree program. Others, like DCC, respect the family for its great contribution during the 1980s and use the top individuals in linebred outcrossing. Nearly all breed-leading cattle today trace multiple times to the Partlow cattle, including Shadow Jubilee, who traces 19 times, and Drag Iron, who traces 33 times to these foundation individuals. Many great cattle would not have been born without the Partlow genetics, including Rio Grande, Royal Reputation ($150,000), Delta Fifi, SDR Candy Cane

($170,000), Gun Man, Top Caliber, Shadow Jubilee ($120,000), Eternal Tari 206, Heavy Hitter, Cowboy Tuff Chex ($165,000), Tempter, The Shadow, Tari Graves FM49, Jamakizm, Clear Point, Mile Marker, Jester, and Hunts Command Respect.

The Butler family is much larger today, allowing an expanded genetic base. The great individuals were preserved, documented, and recorded for registration. Although mistakes were made and wars were fought, Pauline, Blackie, Sam, Milby, Henry, and Uncle Dode had a Texas Longhorn love and appreciation that those of us who enjoy this breed should never forget. I found great joy and honor in helping to locate and document this amazing Texas Longhorn family.

Miss Liberty 10 was 21 years old in this photo. She produced 4 embryo calves after age 21. She was sired by Liberty Lad, a bull purchased by DCC in 1972 from Dan Coates, Sr.

Unlimited is appreciated by many as the leading Butler sire. His progeny are often dark-walnut and have a speckled lacy face that makes them very attractive. His unique color factors come from Maressa.

"I have continuous enjoyment savoring all pasture livestock, with a degree of curiosity which makes some think I am a fool, and some think me far wiser than I am."
～ Darol Dickinson

He's a Dude of Houston

Most people think of themselves as normal. Some are; some aren't. They think like that because they are similar to others in their communities, working in the same businesses, or maybe going to the same churches. Each area has its own individualized sense of "normal." The Jerry Davis family of Perry, GA, was normal in its warm southern environment, but Davis was an extremely interesting and unique man.

The Davis family included several kids; their family activities involved the horse industry. During the **All American Quarter Horse Congress**, I met them at the Million Dollar Stallion Avenue where they were promoting their AQHA champion He's a Dude, a beautiful chestnut roan. He's a Dude was a son of Blondy's Dude. He was smooth and trim, well-trained, easily handled, and a breeze to photograph. Today his progeny have accumulated a total of nearly 4,000 AQHA open-show competition points. The Davises commissioned me to do a 30"x36" oil of their flashy stallion. After the Congress, I rode down to Perry with their caravan of kids, horses, and vehicles.

Jerry Davis was around 60 years old at the time; and like every hard-working person his age, he'd had time to surround himself with what he enjoyed. He'd planned his own "nest" to fit his family.

We arrived late at night. Jerry took me to a guest room in his huge old southern mansion. The bedroom ceiling was about 14' high and had an old original squeaky ceiling fan. Jerry said his dad slept there, but could not sleep well if the ceiling was right down in his face. It was certainly a long way from my face.

After a breakfast that included the local delicacy of hominy grits and fortunately some other normal stuff, Jerry took me around the property. It was a plantation in the full sense. It had been a pecan farm for something over a hundred years and boasted of rows of thousands of pecan trees stretching in every direction. The property also included a huge lake, white fences, horse barns, an arena, and a number of small homes set pretty close together on the far side of the lake.

Jerry's home and property are located in Georgia's Houston County, pronounced "house-ton." This local normal pronunciation is not to be confused with a city in Texas called Houston, pronounced "huues-tin." You are evaluated as a newcomer depending on how you pronounce the county name.

Jerry was raised on this beautiful family farm. He graduated from Georgia Tech, and most thought he would promptly return to the farm. But first he used his engineer training to land a management job as one of the builders of Robins Air Force Base. He joined the 32nd Marine Air Group and served to defend the country in the Pacific theater. After WWII, he returned to the family pecan plantation and started his own family.

My first morning at Davis Plantation, we left the house and loaded into Jerry's bright yellow Checker Taxi. It was a huge, heavy, retired cab with wide seats and four doors; it drove like a tank. These old, worn-out city taxis were easy to buy for scrap-weight price. No one except Jerry would want to use one of these for his main vehicle.

Between the Checker cab's parking spot and up against the house was a row of fig bushes. They were in season, and ripe fruit was falling all over the ground. Nothing is as tasty as fresh, in-season figs. Their sweetness made up for the blah taste of the grits. Being from Colorado, I enjoyed my full fair share of delicious figs.

All people have an inner quasi-animal that prompts them to act in certain ways. Some are very wealthy and work at appearing below normal, or just average and poor. Others don't have a pot or a window to toss it out of and make a shiny public appearance of having great wealth. Jerry was a humble man and found his comfort zone in being normal and fitting in locally. Although he was founder of the First National Bank of Houston County and was quite possibly the wealthiest man in the area, people knew and liked him for his consistently unpretentious appearance. He was kind to small children and was everyone's friend. He was happy to be just a normal person.

Jerry was proud of his historic family plantation—and rightfully so. Several generations of Davises had lived, sweat, hammered out a business, and died here. The plantation produced truckloads of pecans each year. Pecan-picking, shelling, and packaging was a labor-intensive job. As we drove around, Jerry would stop and talk to people who were walking or driving along the roads. They were mostly black men, women, and barefoot children who politely called him "Mista Jerry." I asked him about the 20 or so houses beyond the lake, and he said that was housing for the plantation workers.

Jerry shared with me a family-farm story from his grandfather. The farm had accumulated an ever-increasing number of children too young to work and seniors who couldn't work. His grandfather was very pleased and happy when the day came to free all the slaves; he hoped to then hire back a couple dozen able-bodied men to work the farm. The slaves refused to leave. They came in full force asking to stay and work. They alleged Jerry's grandfather had provided food, clothing, medical care, and housing, and they didn't want to leave that certain provision. Little by little, as generations passed, many did chose to leave, and he was able to hire back good workers. Many of Jerry's current workers have family roots in that farm going back to and before his grandfather's time.

We walked out under some big moss-draped trees into a papershell pecan orchard. Pecans were on the ground as we crunched over them. Jerry would reach down and fill his pockets. He had a way of walking, talking, and rolling two pecans in the palm of his hand. They made a special little cracking sound. Somehow he had the skill to crack one pecan against another, breaking it around the center. Then he'd pull it open without even looking at it—obviously not a new thing for him. He would hand me the nut and then crack another as fast as I could eat what he'd just given me.

Jerry liked pretty-colored horses. (Drab colors are more difficult to market.) With his Quarter Horses, he could never have the flash of an Appaloosa or a Paint, but he got as much brightness as he could register. Many of his horses were palominos, roans, or bright chestnuts. To create valuable colors, Jerry bred horses close to silver, gold, red, blue, black, or white. Livestock-breeding programs like his created horses with strong colors, a lot of décor-white markings, and white manes and tails that were an easy sell. They quickly caught and held a buyer's eye.

Since the purpose of my visit was gathering the material I needed to create an oil painting of Jerry's famous stallion He's a Dude, we drove all around the lake and found many beautiful spots for a background. The hardest part of this painting was the tiny brush strokes for the contrasty roan hair coloring. There were no short cuts; instead I just painted the hair, hairs, and more hairs. The mark of a great portrait is to look exactly like the subject—or perhaps, slightly better. ▷-◁

"If a tree falls in the forest and no one is there to put out a press release—it didn't happen."

～ **Stanley M. Searle**

Working Big Pastures

In 1984, DCC owned or leased pastures in 6 counties of Colorado, from the mountains south of Hartsel to winter-field residue grazing on the Arkansas River Valley around Granada. Most cattle were at the headquarters in El Paso and Pueblo Counties. Regular observation and 100% inventories were imperative. The minute school was out, our kids all had jobs. Chad, our middle son, at age 13 rode pastures all summer with the senior cowman Don Kraven. Don was known to be able to ride quietly through a cow herd and hear every animal breathe. When they were all breathing quietly and accounted for, the horses were loaded for the next pasture, which might be 40 miles away. If a cow needed attention, she was head-and-heeled, treated in the pasture, and the hands were off to the next problem.

There were places, like the Quixote Ranch at Boone, where cows would duck into the brush and be hard to rope. If a cow could be driven out to an open area, it was very important to catch her with the first loop, because a second shot might take a lot of time to get her out again. The less a sick cow was chased, the better it was for her healing. If Chad missed a loop, Kraven had a John Wayne way of "dog cussing" with strong-colorful words of "blue lightning" that explained the stupid, ignorant, dumb, poor judgment, and horrible problem he had caused. Chad worked with all his heart not to miss a loop, but 100% catches under less-than-tender arena conditions were difficult, even for the best.

The Quixote pasture bordered the Arkansas River for about 2 miles. There were fences, but the river had a strong current that made a pretty good fence to contain cattle. It was about a city block wide and got wider when some heavy rain hit the mountains upstream. Every so often, cattle and horses would be swept away by the river and wash ashore at the Quixote. The horse on the left that Kraven was riding came down the river one day and took up residence in one of the ranch

pastures. He didn't have any brands and was a little horse about 13-1 hands tall. The men saddled him up; he wasn't new to the saddle, but needed some wet blankets to be useful. We named him "Shorty." No one ever asked about him or came looking for a slick, no-brand chestnut gelding, so we used him for a dozen or so years. Chad was riding a young gelding sired by Silky Fox.

This photo shows Kraven doctoring an embryo recip cow with Chad holding her stretched. Chad would hear Kraven drive up to the ranch at the crack of dawn. Chad would grab his sack lunch and be gone for the day. Before Kraven showed up, he would have saddled four horses and drunk a pot of strong coffee. Chad would get back home after dark, eat supper, be dirty, sun-tanned, with red eyes, maybe a few rope burns, and be fairly quiet. After supper, he would go out to the barn and practice roping a bucket for an hour or so, then collapse into bed. The next sound he would hear would be the rattle of Kraven's trailer fenders at first light.

Kraven rode about six horses, some young and some solid, well-trained. Each day some were hauled and some rested. Several of the Colorado Springs-area rodeo people would send horses to Kraven for a couple months to get some miles under saddle. Their horses would come back lean, obedient, with tight muscles and a no-nonsense attitude. Some went on to be good barrel and roping horses. Kraven gave them a lot of wet saddle blankets, which will cure most bad habits a horse develops.

Jumping forward more than 30 years, 6 grandkids and 1 great-grandkid are working on the ranch as their formal educations continue. Most work cattle with Honda 4x4's, some sort cattle for breeding pastures, some build fence, some spray-poison weeds, some keep records, some apply vapor ear tags, and life goes on at the ranch. Three of Chad's kids were working today. Chad is a builder in Georgetown, TX, where there are fewer rope burns. The younger kids open gates for me to check pastures on my Ranger Crew. They help get ears up when I am taking photos for the internet cattle-sales inventory.

Kids need a lot of open space, trees, and water. Space helps them sleep good at night. If you box them up in a tight area, they will destroy the property and maybe those around them. Kids need to spend as much time as possible with adults, parents, and grandparents. Adults are not teaching them bad things. They may actually teach them the great honor of work achievement for honest money.

> *"Prepare thy work outside, and make it fit for thyself in the field; and afterwards build thine house."*
> ~ **Proverbs 24:27**

Three Bars, Vail, and Merrick

Three Bars TB—born 1940. For the record, Three Bars TB had 499 progeny that spanned 24 breeding years. He lost some early siring years while at the race tracks. Here is his record: his progeny had 9,077 race starts; 1,584 race wins; 176 stakes wins; progeny race winnings of $3,214,682; 308 Register of Merit; 36 Superior Race Awards; 24 World Championships; Pokey Bar was his leading money winner at $162,632; 1,544 progeny AQHA halter points; 584 performance points; 29 AQHA Champions; 4 Supreme Champions; 5 Superior Halter Awards; and 2,129 AQHA points won by progeny in all divisions. Three Bars was the only Thoroughbred sire to produce AQHA arena cutting winners.

Pokey Bar won the All American Futurity in 1961 and won more money than any other Three Bars son. This photo was taken at Hugh and Sid Huntley's in California.

Most of Three Bar's life he was owned by Sidney H. Vail. Vail stood him to Thoroughbred and Quarter Horse mares in Arizona and Apple Valley, CA. Walter Merrick was the major life-long lover of Three Bars. He leased him for two years and bred him at his Sayre, OK, ranch early in the stallion's career. While at Merrick's, the stallion Midnight Jr. reached over the fence and bit off part of Three Bars' nose. Vail was totally livid by this incompetent treatment of Three Bars, who had a goofy-shaped nostril from then on. In two years at Merrick's, Three Bars was given the right mares he needed to begin a celebrated life-long success, then Vail took him back.

I first saw Three Bars in May of 1959 on my high school senior trip. I forfeited all votes involving selections of California amusements for the one vote for the bus to take me by Apple Valley and see Three Bars. The other classmates were unimpressed, but I was exploding with excitement. Sid Vail, also unimpressed, allowed the load of kids to stand near his stall, and we looked in the door. I got to see the horse—I saw him!

Just months before I was there, right as Vail was about to harvest the highest stud fees in the industry, the golden horse was stolen. There was an all-out search for him coast to coast while someone unknown had a free breeding season's worth of Three Bars foals. I don't know if that mystery was ever solved, but one day just at sunup, Vail looked out his window and saw Three Bars grazing in the driveway. He was back, skinned up and looking worse for wear; but breeding season was over and the famous stallion never told about his nightmare kidnapping adventure.

The years of high-dollar stud fees and selling high-priced progeny made Vail a very wealthy man. Rumor was that he resented Three Bars because of the demands on his life. He had to guard Three Bars. He had to show him to fans. He had to handle the stud with kid gloves for breeding. Three Bars controlled Vail's life. Sid was on the phone all the time to horse people. His 20-plus years owning Three Bars demanded that he have no personal life of his own. But Walter Merrick was always there to help.

The second time I saw Three Bars, I was photographing horses in California. I had taken with me a large original painting of World Champion Running Mare

Straw Flight to help sell commissions. Vail invited me to his huge home, and I brought the painting to show him. Several people did gift paintings of Three Bars. He showed me a new one—another gift. I tried to sell him a Darol Dickinson original—no luck—but I wasn't really surprised. Orren Mixer had also sat in the same room with Vail a few years before and couldn't sell him an original either. Vail liked the free paintings that young artists gave him.

Every year, Walter would make a serious effort to acquire custody of the business of Three Bars. Finally when he was age 25, Walter got him back again.

The third time I saw Three Bars was at Quanah, TX, at the J Bar B on the south side of Texas Highway 287. Walter was breeding him in his 25th year and could turn the crank when booking mares: more than 100 were waiting their turn. I took two mares to breed to Three Bars. The fee was $5,000 each. One was a cash payment; the other $5,000 fee I traded for a 30"x36" oil painting of the horse himself. Walter was a lot easier to deal with than Vail.

When Three Bars and a number of other great stallions were at J Bar B, they had mares in small 2 to 10-acre paddocks each waiting her turn. I had been to a

Three Bars painting—oil on canvas 30"x36"

lot of horse operations, but I noticed that these mares did not have a neck-band identification. In fact, there were no identifications of any kind on any mare. I asked Walter why they did not have these typical visuals, and he said, in his slow western Oklahoma vernacular, "That's kid stuff."

For a half-hour every day, Walter would personally lead Three Bars to a plot where he continued his Three Bars love affair while the horse ate grass. No one other than Walter himself got to lead him, and no other stallion got to eat that little patch of grass.

Three Bars had a large book of celebrity mares every year. His normal pattern was to start the season shooting blanks, so his performance was not for the faint of heart. For about a month each year, his semen quality was on the edge of a heart attack for his owners. Walter was a nervous wreck during those weeks. Then

Each day Walter would lead Three Bars out for some green grass. He knew that he was leading a genetic fortune that many had enjoyed and more great things would come.

about 30 or so days into the season, bang, everything came out okay. At Merrick's, Three Bars had the most professional AI vet in the nation. I recommended that Walter do two breeding seasons, offering a normal spring breeding at the full rate, then "off-season breeding" at a reduced rate for the rest of the year. A few years later, no one would care when Three Bars foals were born. My theory was that the frequent service would keep the stud from having a "blank" time and keep him in some normal physical condition.

Vail never had to worry about Three Bars again. The famous stallion died 2 days shy of his 28th birthday on Merrick's ranch. Walter had prominently hung my painting in his living room—the only oil portrait of Three Bars ever paid for. Even then, the horse bought his own painting. ▷-◁

"Wealth comes from successful individual efforts to please one's fellow man ... that's what competition is all about: 'outpleasing' your competitors to win over the consumers."
 ~ **Walter Williams**

"King"

From childhood, I had read history about the wild and rugged Texas Longhorn breed of cattle. All the cattle people I talked to agreed they were small, crazy, and would not gain well enough to be profitable to any ranch. My fascination with the breed continued despite its reputation. In my heart, I thought surely some really thick, bigger, gentle cattle could be identified in the remaining herds, then bred for the desired traits, and a quality breed could be created. I hunted for such rare individuals that were the most pure, colorful, correct, and gentle. Once this diverse quality was proven, value would follow. In the '60s, registered calves sold for $100–150 and pairs around $300. This was not a good business. Values had to increase, and that could be done only by breeding the best to the best and performance-testing to develop improved genetics. Not only the image, but the "total package" had to be developed. It would take time.

As I got old enough for a driver's license and my business of photography increased, I was able to go to the few Texas Longhorn ranches still fighting the tide—still raising the old breed. In the late '60s, the government herds were thought to be the largest; but the Peeler, Wright, Phillips, YO, Yates, and Butler were actually larger in numbers. Most, if not all, herds were multiple sire, where several bulls were with the cow herd. That always had been a standard practice since open-range days.

In the early '70s, I paid several visits to Battle Island Ranch, the vast estate of the J.G. "Jack" Phillips, Jr., family. When nearly every bull in the industry was under 40" T2T, on my first trip to Battle Island I saw 4 bulls over 40". That was an early treasure find of huge horn. The cattle were also large-framed. The Phillips cattle had been bred for frame, bone, and size, but often were dull, pale, and subtle colors. I went crazy until I could get some Phillips cattle and eventually acquired 6 cows and leased a bull named Texas Ranger JP. That was a start in developing the greatest part of the Texas Longhorn foundation blood—Texas Ranger. His top horn-spread was a fudge over 48" T2T—the largest in the nation.

Jack would not let go of his Texas Ranger daughters. He had some that were low, lateral-horned with tight corkscrew twists. I thought that Miss Texas Ranger 262 was his best and begged him to price her, but he would not budge. He would not sell any of them.

In 1972, I showed up at Battle Island, where T.D. "Terry" Kelsey had arrived ahead of me. He had found out where Texas Ranger came from and was camping at the ranch, eating a lot of Carolyn Phillips' southern chicken-fried steaks. Terry lived near DCC in Colorado and later became one of the greatest sculptors in the world—a persistent, talented person. In fact, about a dozen years later, he sculpted the Texas Gold monument and lent it to TLBAA at no charge. It is one of the world's 10 largest bronze pieces.

We were riding through the pastures, Jack was driving, and all of a sudden an armadillo hopped along in front of us. Terry had never seen one but jumped out of the truck, chased him around trees, in and out of mud, and wouldn't quit until he finally caught him by the tail. After a lengthy chase, the armadillo was exhausted and so was Terry.

"King" was the first Texas Longhorn to weigh over a ton and measure 65" tip-to-tip. He moved the bar forward with thickness and the total package all at the same time. Those who used him are glad they did.

During this pasture tour, I saw Miss Texas Ranger 262 again. And once again I asked Jack her price, but it was too late—Terry had bought her the day before. Sometimes superior genetics are nearly impossible to get, and getting them takes extra time. I waited. Then, years later in 1980 at the Texas Ranger Memorial Sale, of which Terry was an organizer, she was consigned. It was my chance at long last to get her. Lowell Goemmer also wanted her and bid me up to $3,500—but I got her. She was bred to Royal Mounty by Texas Lin and had a red bull in the spring of 1981. I had some special dignified names picked out for special bulls: Classic, Impressive, Emperor, Monarch, Conquistador, and King. We named this red bull "King" spelled with quotation marks. We had already used most of the other reserved names. "King" was red, and red was not popular. It was not a high-dollar color. Before "King," we had introduced Classic, who was white, and white was not popular either. There were hundreds of white cattle. We promoted "King" as the antidote—the white corrector, the fixer—to reduce the white. Not only could he kill the plain whites, he weighed over a ton and had more horn than the famous horn leader, Classic. He made for a beautiful blend, putting a lot more size in this new generation—the Phillips size. He was the first mid-60" horned bull, with a whopping over-a-ton weight.

"King" was bred natural to over 100 cows a year by DCC for 5 years. He sired Whelming King, who was Ben Gravett's top sire for years; Queen of Kings, TLBAA World Champion; Kingly Blend, the grand dam of Super Bowl; King's Lin JWT, the dam of famous clone cow Overlyn owned by Burton/Stockton; and he sired the big guy Zhivago. When Zhivago came of service age, "King" had completely replaced himself.

I offered to sell "King" to Blackie Graves, as his herd was almost all white and on the small side. I knew this would be a blend from heaven for Blackie, but he would not cough up the money. I offered "King" to Betty Lamb, who also could have really enjoyed the blend in her herd. Carroll and Paula Shores, who were newer in the business and really working to get a quality herd, bought "King." He became the king of their herd at Ben Arnold, TX. They used him several years and raised some wonderful progeny. Joe Assad used him one season, and then he was purchased by El Coyote Ranch at Kingsville, TX. El Coyote used "King" then sold him at auction, and Mickey Wood bought him at age 14 for $8,700. That was a big price for a bull that age in 1995. In the '90s, people doubted that any bull of any breed would bring that price at that age.

"King" was a significant step up the ladder for the Texas Longhorn breed, not just to grow horn, but to be competitive as internationally profitable beef cattle. His genetics are found in over a dozen countries.

"King" may be spotted in pedigrees of RZ Temptress, Clear Point, TCC Shutterbug, Becca II, Jester, Pretty Lady, Zhivago, Clear Win, PCC Rim Rock, Kinetic Motion of Stars, Field of Pearls, Drag Iron, Over Kill, Tempter, etc. Remember "King" is always correctly spelled with quotation marks. Any other King is not the "King." ▷-ᴅ

Zhivago was taken to the Colorado Springs City park, The Garden of the Gods, for photos. Normally a bull can be photographed on a city park or national park for about a half-hour before being forced off by federales. Zhivago added substance, beef, and thickness to the breed with leading horn genetics. He weighed 2200 pounds.

> *"When you can direct your passion into a business opportunity, you have then found a huge piece of the puzzle to potential success."*
>
> ~ **Dale Hummel**

Splash Bar and Michael Mulberger

Splash Bar was born in 1963, bred by Hank Wiescamp. His dam was Skip Irish by Skipper's King. He was sired by Bar Mount, who was out of Question Mount by Gold Mount and out of Red Bird Shoemaker, who was by Nick out of Plaudette.

In 1969, Linda and I were guests of L.R. and Mary Ann Spires at their home in Ruidoso, NM, during the All-American Futurity Race events. I received a call from Michael Mulberger in Wisconsin who badly wanted a stallion photographed. He was winning a lot of shows, but the photos he was getting showed that the horse had flaws. His horse, Splash Bar, was dynamic and came into the show ring with a first impression of greatness that assured the Grand Championship. If a judge looked real close, he might reconsider, but most did not change their mind. He won a lot. He was bold and eye-catching.

Michael said whatever it costs, get on a plane to Milwaukee. He picked me up at the airport driving a white half-ton pickup. Michael was about 25, a very sharp, handsome fellow, with a sign on the truck door: Shiloh Quarter Horse Farm. I asked him why he named his farm Shiloh, and he said in the Bible *Shiloh* means "abundant peace." He said that when he had been saved and became a Christian, a peace had been given him like never before. He quoted a lot of Bible verses about being saved and was excited about his new rewarding life. He apologized that he needed to stop by the brewery and sign some papers before we could take the photos. (Brewery—Christian, interesting)

He pulled up to the first parking spot by the front door of Miller Brewery, where a sign said MULBERGER. I waited in the truck. In about a half-hour, he returned and apologized some more. He said his mother was Lorraine Mulberger, whose maiden name was Miller. Mrs. Mulberger had inherited controlling interest in

Miller Brewery, and Michael was her one son. He gave all the details of how he had been saved and was reading his Bible—his life was taking a major direction change. While reading the Bible, it dawned on him that the growing success of his family was due to the increased solicitation of millions of people who would consume more and more alcohol.

As a result of his conversion, Michael could not with a clear conscience be in the beer business. He had chosen to do a complete stock exchange for $36,000,000 of shares in Dow Chemical and other good and valuable considerations. On that very day when he walked out of the family building, which was nearly a half-mile long, he was no longer in the beer business.

He apologized more about the brewery as we drove to Shiloh horse farm. His farm was perfectly manicured, not a large place, with a beautiful barn, as I

Miss Sun Oil was owned by Michael Mulberger. She was wonderfully photogenic from every angle. What a striking profile years before Photoshop.

It was a cloudy day and hard to get the muscles to pop out. This is the photo Michael wanted to promote his favorite stallion, whatever it cost—Michael never complained about price.

recall about a 6-stall. An old fellow slowly walked out of the barn to greet us. He took care of the stalls and had a Norwegian name that I don't remember and I probably couldn't spell. He had broken English and a kind, warm smile. Michael wanted me to meet this special fellow. On the way to the ranch, he told me that the "barn guy" had helped him understand how to have his sins forgiven and become a Christian. He was eternally grateful to him. The barn guy had about a third-grade education but knew the Bible cover to cover. As I stood in the presence of a 25-year-old 36 times a millionaire and a barn guy, the magnitude of what had happened between these two men was soul-stirring. These two very different people showed a closeness and respect that only the Lord Jesus could ever dream of gluing together.

Splash Bar was led out to a grass area. The light was hazy, so we had to wait most of the day for brighter light until the shoot could be finished. He was an outstanding stallion and easy to photograph, with great poses from every angle. The barn guy had him so perfectly trained he stood motionless. We had a lot of time to talk before I had to get to the plane, so we sat on the grass, where Michael and his dear friend shared their favorite Bible verses with me. Of course I enjoyed quoting a few precious verses that were close to my heart.

Michael enjoyed showing horses and acquired several more outstanding competitors, including the beautiful many-times champions Miss Sun Oil and Cahuenga Kid. He had a keen eye for correct horse-flesh.

Michael moved to Arizona and went into full-time Christian work. I was told someone met him street preaching. He was using his assets for the homeless and doing missionary work helping others to become Christians. While traveling the country snapping photos, I was amazed to make contact with some awesome people like Michael Mulberger. He was a prince, a man with solid convictions, the young man of Shiloh. ▷-D

Cahuenga Kid by Don Bar. He was AAA, SI 90, AQHA Champion, Superior Halter, 202 AQHA points, and was shown by Michael Mulberger. He represents the period when a banding with some close-up Thoroughbred blood was replacing the bulldog-type Quarter Horse in the championship show ring.

"There are many burning bushes. Those that see, remove their shoes and bow; everyone else just keeps picking berries."
⁓ **Mitch Zajac**

Starting the Nevada Herd

One spring, a young man called with a lot of questions. He wanted to start a Texas Longhorn herd in Nevada, where he was starting to raise a family. He had been reading the ranch **www.texaslonghorn.com** site for months and was getting ready to pull the trigger. The DCC ranch site has over 400 articles, with illustrations, about ranching, cattle management, marketing, etc. Entry-level producers can get mind-boggling information if they bury themselves in this pile of data and read the "how to" articles.

Let's call the young man Lance, because I don't want to hurt anyone's feelings. Lance would call every week or so with long conversations about everything involving cattle profit. He called all summer and fall. There are people who have to know every part of an investment at the beginning. Although I encourage gathering a lot of information, other things will happen—it takes time to identify management problems and solutions. Doing business is the fastest way to learn. It is totally impossible to identify everything that can go wrong, then fix it before it breaks. Some can't stand the thought of a failed plan. Lance called several more times in the fall and, over time, committed to buy three of our most economical heifers. He had dozens of questions about every part of the pedigrees, colors, disposition, and payments. DCC requires a purchase deposit to hold cattle off the market. He assured me a deposit would come. People who are slow pay, or don't pay when they say they want to buy, aren't necessarily bad people; often they just don't have the money.

Lance wanted me to deliver the cattle to him. He was in the middle of nowhere, and sending a trailer with three little heifers was a problem. I try to make deals work. I offered to make calls to past clients in that big western world to gang together a load. If I could sell a whole trailer load, the cost for hauling per animal would be much less. As weeks went by, I sold 2 bulls going to California, 3 young bulls to go up around Salt Lake and 4 cows around St. George, UT. Lance decided he could borrow a trailer, come get his heifers, and do the hauling services for my other buyers. He would get paid to deliver the other cattle and haul his own cattle free—a good plan. Deal done. Problem solved.

I put Lance's ride together—a ready-made deal. I had collected from the other buyers, but nothing from Lance. He messed around and finally said he wanted to see the cattle in person before he wrote the check. By this time it, was the dead of winter in Ohio's Appalachian foothills.

Lance showed up during a snow storm and got stuck about a mile from the ranch. Our DCC men met him with a big tractor and pulled him up to the loading corrals, where his calves and the other cattle were all penned, ready to load. I walked around the pen with Lance answering questions, and finally went to the car to thaw out after nearly an hour of multiple questions. He continued to look at the calves.

Finally, it was evident why he was stalling. When he and his father-in-law, a Bishop in the Church, got into the car, he said he had spent about half his money on fuel driving to Ohio and did not have the money to buy the calves. He needed the rest of the money for fuel going home and doing the deliveries. Would I give him the calves, and he would pay after he collected from the other buyers for hauling? He said he could do the deliveries but did not have the money to deliver and also buy the calves.

A seldom-broken rule at DCC is that when cattle get on a buyer's trailer, they are paid for. Repossessing stolen cattle in Nevada is not an easy thing.

After six months of planning and hours of phone calls, I could not believe it had come to this. I had worked to make this financially good for Lance. I had worked to put this trip together for him. Here we sat, with snow blowing and windshield wipers snapping back and forth.

At this time the father-in-law Bishop began to encourage me about Lance. Lance was young, had been married just three years, had three young children, was faithful in church, and was a hard worker at a local chicken farm. He was honest. You could trust him for this small amount of money. He would keep his word. He was a good father. The sales pitch was getting more desperately compelling.

As we sat and watched the wiper blades, I wanted this deal to work for everyone. The wipers, I realized, were the only things doing what they were supposed to do.

I made a recommendation. I offered to load all the cattle for everyone, with a check from the Bishop for the three heifers. He knew Lance better than anyone. He could write DCC a check, then when Lance got the whole deal done and had collected for hauling, Lance could pay the Bishop back, and all would be well. I was comfortable with that idea. In fact, I was pleased with myself to come up with the idea. No one said anything for several minutes, then the Bishop said, "No. I won't do it." Then it was obvious—he knew a lot more things about Lance than I did.

It was with some concern that we loaded only the Utah and California purchases into Lance's borrowed trailer. The cattle were delivered as promised, and our clients paid Lance for the hauling. He even rested the California cattle at his Nevada place before taking them the final leg of the trip. It went amazingly well.

Before Lance and the Bishop left, he assured me he would deliver the cattle, then come back to Ohio and get the calves. I'm still waiting—that was last winter.

It was snowing and cold, but our men gathered the cattle, got Lance loaded, and dragged his rig to the freeway with a big tractor.

"There is nothing so useless as doing efficiently that which should not be done at all."

~ Peter Drucker

Frank Doherty and Doherty 698

Doherty 698, a very valuable Texas Longhorn foundation cow, is most famous for being the dam of two legendary sires, Phenomenon and Overwhelmer. These two bulls alone created millions of dollars of value in the Texas Longhorn industry. Doherty 698 was genetically ahead of her time. You would never guess that her interesting story traces back to a zoo in Tennessee and a coal company in Ohio.

In the early '70s, Memphis Zoo acquired a specimen pen of Texas Longhorn cattle. These zoo cattle were from the Wright herd of Robstown, TX. In those days, it was not unusual to find Texas Longhorn cattle in most big zoos. A cow and calf were valued around only $250–400. The value of the breed was considered to be only in their history and in the public's delight in their twisty horns. No one was exploring to find their deeper value or to preserve their best genetic characteristics.

During the same period, Federal law required coal companies to reclaim their strip-mined lands. This land had been drastically molested, so land owners searched for ideal ways to restore it to profitable use. The huge Consol Coal Company of Cadiz, OH, owned over 60,000 acres of rugged reclaimed-mine grassland. It had planted 10 experimental grass mixes and thorny wild brush on its Appalachian hills and came to the conclusion that Texas Longhorns were the ideal critter for grazing them. George Cobb, herd manager for Consol, was accumulating a sizable herd for the coal company. He had bought 4 adult herd sires from Frank Doherty of Fort Scott, KS. One Doherty bull was Señor Mulege. George also read an ad about Memphis Zoo having 5 Texas Longhorn cows and calves for sale. He purchased the zoo cattle and the cow Wright 489. Consol's herd grew to be one of the largest Texas Longhorn herds in the nation.

Frank Doherty kept in touch with George following Cobb's purchase and paid attention to the growth of Consol's herd. Some of those cattle struck his fancy, so he bought several in the late '70s, building much of his foundation herd using cattle that originally lived on the strip-mined hills of Ohio. One of those was a Consol heifer born on 3-15-77, sired by Señor Mulege, originally owned by Doherty, and out of Wright 489, a cow acquired from the Memphis Zoo. Doherty named her Doherty 698.

Frank and his wife, Brownie, dreaded naming all their registered Texas Longhorn cattle. He found some easy ways to shorten the turmoil of finding names. One method meant putting an ID brand number on every critter and naming them Doherty plus their ID number. One year, he named them after states—Miss Arizona, Miss Kansas, Miss Ohio, etc. When he ran out of states, he named progeny Miss Phoenix, Miss Dodge City, Miss Toledo, etc.

DCC wanted to locate genetics to add size and horn to the TL breed. But this was a challenging task as most registered TL cows were only 600–800 lbs. Doherty

Wright 489 was bred and raised by Wright Materials at Robstown, TX, went to Memphis Zoo, then to Consol at Cadiz, OH, and on to the Frank Doherty herd. She is the dam of Doherty 698.

698's Wright-bred dam had impressive size for the time. Her genetics resulted in Doherty 698 reaching a top weight of 1,370 lbs. This was very large for the breed. She had horns 44½" T2T. A wild-shaped shield in her upper forehead was a distinguishing beauty mark found on many of her progeny. I tried every way possible to buy Doherty 698 from Frank, but it never happened. Fortunately that did not prevent DCC's being able to establish her genetics in Texas Longhorn history.

Non-surgical embryo transfer was being perfected by 1979. Quickly DCC developed an extensive embryo-transfer program in Colorado. A specimen herd of 86 of the best of the best Texas Longhorn cows to be donors was collected from leading producers all over the nation. DCC did not have the money to buy all these great cows, so most were acquired on a 50/50 partnership deal. We did all the embryo work and paid the vet bills, and the owners of the cows split the resulting pregnant recips with DCC. To get Doherty 698 into the program, Frank demanded 2/3 of her pregnancies. In the 50/50 deal, he also wanted several other

Doherty 698 was the top cow of the breed in 1980 when she started in the embryo program. She is age 4 in this photo.

cows that we did not want since they did not have enough value. The deal was struck. While DCC got 1/2 of all the other 85 donor cows' calves, it got only 1/3 of the Doherty 698 calves. As he wanted, Frank got the other 2/3 of the Doherty 698 calves. He was a ruthless negotiator.

Doherty 698, Measles, Rose Red, Maressa, and Ranger's Measles produced the most valued pregnancies of the 86 donors. Many of the pregnant recips were pre-sold for $10,000–$12,000 each. Those 4 cows produced over 110 embryo calves. Although Frank got 2/3 of the Doherty 698 pregnant recips, DCC got the recips pregnant with Phenomenon and Overwhelmer. There were other full brothers to both of these bulls that were possibly better, but they were sold to producers not experienced in breeding and marketing. They just melted into the vast Texas Longhorn world and evaporated away.

Doherty 698 produced 44 embryo calves in a little over 2 years. The industry did not have the great bulls in the early '80s, nor was there any track record of her previous progeny. She was flushed to 11 different bulls, always hunting the right matings. She was so fertile, all 44 were produced by the time her first flushes were weaned. They were by Texas Ranger, Don Abraham, Bold Ruler, Superior, Cowcatcher, Ranger's Ranch Hand, Rural Delivery, Measles Super Ranger, King, Classic, and Classic Mac. With few exceptions, she produced the best progeny ever for each of those bulls. She singlehandedly took the breed up to the next level back then, totally a result of embryo transfer.

As Doherty 698's great production became obvious, Frank and I decided to syndicate her. We offered 6 interests for $20,000 each. Wayne Rumley, who was the main owner in Tri-W Ranch of Tulsa, had bought a half-dozen or so Doherty embryo calves. He loved this cow. The syndicate was promptly sold out to Wayne Rumley, Bill Atherton, Wes Hill, and Lane Mayberry for $120,000 total. DCC continued to flush her two more times for the new syndicate. Later, in her upper teens, it was reported that her tip-to-tip spread grew to nearly 80".

Although few today got to see this early dominant cow, almost everyone owns cattle who trace back to her. She is found in the pedigrees of 3S Danica ($380,000), Phenomenon, Kinetic Motion of Stars, Shadow Jubilee ($120,000), M Arrow Cha-Ching, Clear Point, RM Miss Kitty, Overwhelmer, Emperor, Overlyn,

In the embryo partnership, Frank Doherty always wanted the big recip cows. I got the small ones. This is a 2-year-old Red Angus x Longhorn cross. She was a wonderful cow, raising Overwhelmer to be nearly as big as she was at weaning.

Auze EOT, Working Woman, Clear Win, Rodeo Max, Tempter, Drag Iron, Lazy J Bowhunter, Over Kill, The Shadow, Eternal Diamond, Winning Honor, and a cast of thousands. Her impact is evidenced by appearing 18,554 times in the TLBAA *Horns* database.

This amazing adventure changed the Texas Longhorn industry forever—and it all started right next to the big elephant display at Memphis Zoo. ▷-D

> *"There is no greater joy than in sharing a passion with those who are dedicated to excellence, who are like minded in their pursuit."*
> ～ **Cathy Hawke**

The Pretty Penny Ranch

Modern horse breeders evolved from ranchers, rodeo athletes, trail-hand cowboys, Hollywood cast members, and old historic pioneers. This out-cropping blend is found in colorful packages, some with money, some dirt poor, yet all have a special love for great horses. The challenge of raising or owning still greater horses eternally keeps the adrenaline flowing in their veins.

I first met Charlie Dees at the Colorado State Fair in the early '70s. He had acquired from Hank Wiescamp a big, stout bay stallion named Skip's Alibi and was campaigning him at halter and different AQHA performance events. Charlie was a great steer-roper, competitive at the highest levels. I was at the fair working to make a living photographing horses and taking orders for oil portraits of horses their owners could afford to immortalize. Photography for promotion was much faster than doing portraits. I completed about 12 oil paintings a year but photographed hundreds of horses and cattle in the same time.

Charlie called me about doing a photoshoot at the Pretty Penny, his place in Arizona. He had never been able to get the exact correct photo he wanted of Skip's Alibi. He had several other clients lined up for me to photograph their horses. The day came; I flew to Phoenix and went out to the Pretty Penny Ranch on Thunderbird Road in Scottsdale. It was about 40 acres in the edge of the desert with nice houses not far away and a crisp view of Camel Back Mountain in the background.

Charlie was a very successful businessman. He had taken horse training, managing, stallion service, and boarding to a very high level. He had designed the Pretty Penny with every detail of perfection. He charged the highest price per day to board valuable horses—more than any stable in the area—but clients got what they paid for. There were 2 or 3 circle corrals and a couple arenas with clients working horses in every one. The Pretty Penny had a nice office for Charlie and big barn wings off in several directions with private stalls on both sides. Owners could come train their horses, store their tack safely, use the Pretty Penny as a home base, and buy, sell, or do the horse business at whatever level they chose.

It got hot some days, but early mornings were often the best part of the day. Charlie had a private deep-well irrigation system that kicked on at exactly 5:00 AM with high-pressure rainbird sprinklers every 60'. The whole place—landscaping, arenas, buildings, and parking lots—all received a gentle irrigation for a half-hour. When the clients showed up to ride in the early dawn hours, the desert dust was gone and a pristine feel of fresh rain prevailed. The Pretty Penny was, in the '70s, very professional. It offered anything the clients enjoyed or needed, from a photographer, special vet care, security for valuable horses, and beautiful flowers growing for décor, to a full line of equipment including horse trailers of every size you could have wanted.

And the Pretty Penny Ranch had bathrooms!

As I traveled from one horse outfit to another, I was always fascinated by what made one place highly successful and others just run-down dumps. Charlie said the secret was his bathrooms. With a lot of female clientèle, he had learned to have large, very clean bathrooms with expensive tile, the fanciest fixtures, great lights, and some relaxing lounge areas. Over the years, he had found that one of the first things people checked out, when deciding to spend a lot of time with their horses, was the bathrooms. They gave people a quick image of the detail in every part of the Pretty Penny Ranch. This was a classy place where high-rollers were totally comfortable and could feel clean in their élite style. There was a psychology to everything Charlie did. He was a student, yet always coachable.

I got to the Pretty Penny with enough time to do one photoshoot before sunset. Skip's Alibi was slick, well-trained, and ready. We hauled him over to a pretty clean lawn near a large factory and used the low evening sun to illuminate his Wiescamp muscle and capture his profile. A well-disciplined horse will take about an hour to photograph. Unruly ones may take more than one photo session of several hours each. This is the photo selected for the postcard Charlie wanted

Skip's Alibi was an easy photo model. This is the postcard photo with Camel Back Mountain in the back.

to use for promotion. When photographing for promotion, the horse owner pays the bill, but the trainer or showman normally handles the photoshoot. I always had the handlers stand at an angle so they would be cropped out of the photo. Handlers come and go, but one great photo will often last the horse's whole life. Today, most handlers jump right under the horse's jaw so they can't be cropped out. It is more important for a trainer to be standing front and center for his personal promotion rather than the stallion's. If there are feelings over it, I would take some of each. Let the trainer have his and the owner have what he wants—everybody is happy. Sometimes a good shot can be done with the same pose just by moving the handler farther back or up closer to the horse.

The Pretty Penny was equipped with a horse-photo area for clients, with a slight rise in the middle and a classy desert stone-wall background. The site was planned with the light at a correct angle. A lot of great horses were photographed there.

After a long day of work was done, Charlie would often take me out to a good Mexican restaurant. With the sweat and labor of the livestock industry come some legendary stories. Livestock people aren't government employees and often make up their own rules as they go. It is about hopefully buying and selling stock at a profit and making do. Over a plate of chile rellenos, Charlie recounted this story:

One hot summer day, he was hauling a load of horses to a show and noticed in his mirror that a tire had blown on the trailer. He and some of the horse people got out and changed it as fast as possible. Charlie said it was hot, too hot to handle. Rather than toss the tire into the trailer nose, they just left it beside the road, planning to pick it up on the way back home. As they returned from the horse show, they were going down Yarnell Hill, a 1,300' drop, on Arizona Highway 89, and to their surprise, below them was a huge desert fire with smoldering smoke down in Yavapai County. The dry grass and cactus were on fire for miles, with fire trucks and emergency people everywhere. As the returning Pretty Penny show team proceeded, they planned to pick up the wheel. While they were trying to find it, a highway patrolman in the road flagged them down. Charlie asked what the problem was. The officer said some idiot tossed out a hot rim in the dry grass and set the whole desert on fire. Charlie said, "Wonder what fool would do an ignorant thing like that?" From then on, no one wanted to get out and claim the wheel.

Charlie had been very successful. He had surrounded himself with a beautiful facility, quality employees, and clients. Everything was planned great; however, he seemed unhappy. Every detail was organized to perfection, but the big challenges in life just weren't there. All the problems had been solved. He thought about it for a long time and decided he was just not thankful enough for his blessings—his mind was messed up. After months of mental anguish, he put his huge Scottsdale home with six bedrooms up for rent and moved out. He rented it to a well-dressed lady who appeared to be a class-act. She paid a lot for rent, always right on time every month.

After getting his house rented, Charlie drove one of the old Pretty Penny pickups to southern Arizona and hired out picking cotton. It was obvious that he was not the normal cotton picker from south of the border, yet he was a cheerful fellow and got along well with the other "pickers." He said by sundown the first night his back was a wreck and his hands were scratched and worn. Day after day, his back got worse. His legs cramped. His hands were torn up. The hot Arizona sun beating down was a far cry from his air-conditioned office. Day after day the field owner weighed Charlie's cotton sack and encouraged him to call it quits. Charlie couldn't pick as fast as the Mexicans, but he worked from daylight until dark, putting in the hours and weighing more cotton sacks.

Finally after weeks, the picking season ended, and he felt he had taught himself the lesson he needed. He was ready to go home and enjoy his other life with a better, optimistic attitude. He was ready to give the lady notice and move back into his house and his own comfortable bed again.

When Charlie arrived home late one evening in the old pickup, his house was surrounded by cop cars, TV cameras, and all kinds of excitement. He had arrived

Charlie Dees holds his top Thoroughbred stallion on the special photo mound at the Pretty Penny Ranch. This photo was also used for promotion and a series of postcards.

just in time to see his house raided and a brothel business broken up—in his beautiful home. He said the damage to the house was more than double the rent he had received. While watching from a safe distance, he recognized two men and one woman he knew from Scottsdale as they were walked out with handcuffs. He wisely decided just to drive on and come back later. He was ready to repair his house, be thankful for his many blessings, and get back to the Pretty Penny and his air-conditioned office.

Once things quieted down, Charlie returned to his beloved Pretty Penny Ranch, confident he would soon have things running smoothly again. Then he checked the women's bathroom. It was completely trashed—his challenges were all multiplied. But he was home and happy.

"Profit is a signal that valuable services are being rendered to people on a voluntary basis."

~ **Lew Rockwell**

Hank Wiescamp — The Legend

Henry John Wiescamp, "Hank," was born in 1906 to a farm family in Holland, NE. As a young man, he migrated over the serpentine La Veta Pass, beyond the Sangre de Cristo Mountain Range, to the high-elevation desert valley and the city of Alamosa, CO. He died in 1997 at 91 years. His wife, Freda, had died before him. When they were young, he had promised Freda that he would build her the biggest gold-brick house in Alamosa, and he did. I never saw her out of the house. I think she was sort of a recluse. They had several kids, and a number of tragic deaths happened in their family. Hank was King of the Mountain, but others around him seemed not so fortunate.

Me and Hank. This was by the back door of his gold-brick house. Inside this door was his office, where millions of dollars of deals were negotiated: just a humble office with 1 desk, 2 chairs, and trophies to the roof, some 50 years old.

Hank's son Grant lives about seven miles southwest of Alamosa and is a sharp, friendly farmer who raises truckloads of beautiful, high-dollar, irrigated alfalfa hay and markets to the Clovis, NM, dairy industry. He smiles when people talk of his legendary father, but to replace or be another Hank would be totally impossible for Grant or anyone else.

This is me on the left and Hank, by his office/home door.

As Hank got older, he had fewer obligations, less travel, not so many horse shows. He hardly had time for normal people when he was in his 60-year prime, but after age 80 he had time to visit. I would call him, make sure he was available, and drive to Alamosa to spend the day with him. I asked him about linebreeding—when to know you had gone too far, questions like that. Sometimes he would give a detailed answer that made all the sense in the world, and other times he would shut me out with one of his famous homemade clichés.

During 25 years of dealing with Hank, I photographed most of his best horses. He ordered portraits of Skipper W (3 of them, including a head-only India-ink drawing), Skipper's King, Skipper's Smoke, and Skip's Reward.

Hank's most famous trainers were Jack Kyle, Leroy Webb, and Margarett Orr. When Hank sent them to shows, he got so nervous and tense he could not stand to watch in person. His whole life was bound up in the genetics, in every part of the anatomy judges were looking at. Hank's adrenaline was pumping. He had to know, but he couldn't stand to watch. He would get mad at his people if they did not call him immediately with show results in detail. His horses had been to hundreds of shows, he had hundreds of mares in the pastures, but every show placing was a huge deal to him.

To the public, the Alamosa Livestock Auction was Hank's visible business. He held a big cattle auction one day a week. The night before sale day, he would call 20 ranchers and ask what consignments they were bringing, and then he would call 10 buyers to alert them about these "great" cattle they would not want to miss. He could tell you the market after he called these clients. It was something he had a sense for. People knew he would have a full pen of cattle, and they knew he would have an equal number of buyers. He did it every week.

Hank was frugal. He also used the sale ring for breeding mares. He would sit in the front-row seats, and his men would bring Skipper W or Skipper's King and certain mares for hand-breeding. Although most of his stallions were pasture-bred, certain studs were bred under his visual control. I never saw him keeping records. It was in his head. Hank loathed selling his best genetics saying, "What kind of house could a builder build if he kept selling his bricks?"

During Hank's early years, things were tough. He was cheap. He called himself a Dutchman. Little by little he bought land and over years became an owner of the Alamosa National Bank, where the ranchers banked. He knew when ranchers needed money to buy cattle. He knew what their cattle sold for and how much they paid against their mortgage. He knew when a ranch was struggling. Somehow he would befriend struggling ranchers; and when the day came the bank was putting the clamps on someone, they would go to Hank. Hank would offer to buy their ranch. He never paid a big price; he bought at low prices. But the thing about going to Hank, he would make an offer and pay cash tomorrow.

As a result, he became one of the major landowners in the San Luis Valley. He put together the best alfalfa farms with the highest-flowing artesian wells in the valley. After buying a ranch, he often hired the family to work it for him, living on the home ranch the rest of their lives.

Hank was a 24-hour-a-day business guy. With him, it was all about business. His good friend Cecil Dobbin, who owned the great Appaloosa stallion Bright Eyes Brother, was good at frequent advice. He scolded Hank for spending so much time with the auction, cattle, and horse business. He alleged Hank was ignoring his family and told him he should do something with his boys—take them fishing, hunting, or some kind of sport. Hank thought it over a month or so and decided Cecil's advice was

Skip's Count was by Skipper's King. Hank would not sell one of his best to locals. He would splatter his best far, far away so he did not have to show against them. Dad and I had to go to California to get Skip's Count. While he beat all of Hank's best stallions in the Rocky Mountain/Colorado area, Hank never asked about or made any complimentary comment about him. We had messed up his master plan.

Hank held Skipper's King while I took this photo with my little 35mm in the early '60s. At the time, I wasn't a very good photographer, and Hank couldn't make him stand still. Hank loved this horse. He was extreme in every way. He was tall, long-necked, had huge jaws and muscles, and yet was trim in the right places. He had the things Hank worked to get many years of his life. He could never be bought, nor was he ever priced.

good. He talked to his boys and obviously shocked them. He was a New Dad. They were excited. Hank decided to take them pheasant hunting—ringnecked pheasants were all over his farms. He bought new shotguns, shells, camouflage vests, everything to become a hunting family. The boys were amazed at this new idea of Hank Wiescamp, the outdoorsman. Freda, his wife, wasn't looking forward to picking feathers, gutting, and cooking pheasants, but she, too, liked this new family-man thing.

The day came when Hank was determined to get away from the business. A thrill was in in the air. He and his boys drove down south of Alamosa to one of his farms, got all their hunting stuff ready, and walked in line across a grassy sagebrush pasture. Hank was sort of a kid again and happy about this change in his normal direction. There was a lot of shooting and blasting, yet not a lot of feathers. Eventually, most everyone shot a bird. The noise of the hunt could be heard clear to Alamosa. Then Hank heard another noise—that of a pickup with a tall antenna coming his way across the sandy pasture with a Colorado State game warden. The officer asked if Hank knew when pheasant season was. Hank did not. He asked if this amateur hunting party had standard Colorado small-game licenses. No one did. The warden confiscated the birds and wrote Hank and his boys all hefty violation fines. No one ever heard about any more hunting among the Wiescamp family. One good thing, at least: if it had been a legal season, they would not have gone over the bag limit.

It was said Hank could look at an auction crowd and predict the sale average before he dropped the hammer on the first horse. I asked him how he did that, and he said, "Just a feeling you get." Once I asked about evaluating young stock to know what they would be at maturity and he said, "Some people just have a trained eye for it." He shut me out. Once I asked him about banding pedigrees and he said, "Darol, you really should lose some weight."

I asked him about inbreeding, which he had mastered to the fullest. He said the first sign of overdoing it was the brain, just like people—they get goofy. The second was non-visible defects which may be infertility or unseen internal problems. The third and most drastic was outward visual defects. He knew.

Hank had over 50 stud pens all made from pine poles native to the area. Quarter Horses were in most, with Appaloosa and Paints off in the back. Hank's nutrition was simple: pure alfalfa hay for a lifetime, never seeing the bottom of the pine-pole hay racks. In order to continue the unique linebred foundation Old Fred program, Hank used nearly all of the 50 stallions held in ready reserve. There might be only one or two mares bred to a certain stud, but there would be a reason in Hank's mind for that mating. Everyone knew which stallions Hank liked best as they were in the front pens just west of the gold-brick house. Every gate had a big lock. Every pen had constant-flow artesian water. He could look out the window and see them.

We were talking about raising Skipper's King and Skipper W, and he said about getting it all together, "It is hard to get every squirrel up the same tree."

He described the hind leg of Skipper's King, who he called the king of the Skippers, "Hind leg like a Siberian wash woman." His goal was to raise horses with a Thoroughbred frame and Quarter Horse muscle. He fully mastered the muscle part, with beautiful type and flash, but was always hunting to improve the extreme frame. He loved speed horses and the athletic look of speed.

In 1993 we moved DCC to the Ohio River Valley grasslands with our Texas Longhorn breeding efforts, so I was a lot farther away from Alamosa. Hank was amazed at some of the

Nick W was never a front-row Wiescamp stallion. He was always about a dozen pens from the top. This painting was a 20"x24" showing the Quarter Circle I brand, which until his very senior years Hank placed on the left jaw of every horse he felt worthy.

prices I was getting for Longhorns. During the '80s, I had sold several for over $100,000. He was amazed that this market had been developed. Although I never could sell him Texas Longhorn cattle, it was like he finally considered me a worthy marketer and would talk shop in a business-friendly way. He had great respect for a private-treaty marketing person—someone who could do a deal. Although he cried a lot of sales during his life, his own high-dollar horses all sold privately. I was honored to have known Hank and spent many days with him. I asked the hard questions—some he answered.

Henry J. Wiescamp died at age 91 in 1997, and on June 15, 1998, his family held a dispersal sale of estate horses. Genetics that Hank would never price were sold to the highest bidders. Skipper Zane, a 13-times linebred Skipper W chestnut stallion, topped the sale at $400,000. Many of the range broodmares were not halter broke, so new buyers were full of surprises at load-out. There were 821 potential buyers registered to bid, with bank approvals, representing 42 states and 3 foreign countries. The sale total was $3,072,700.00 on 264 horses, an average of $11,639 each. My dad attended the sale and talked about it the rest of his life. It was a horse-world social gathering. Most do not recall any other sale of this electrifying magnitude. Hank Weiscamp was the only man ever inducted into three breed association halls of fame: Quarter Horse, Appaloosa, and Palomino. ▷-◁

Skipa Star was sired by Skipper's Lad by Skipper W. Hank searched his whole life for the perfect outcross blood for his intense linebred genetics. Skipa Star was a wonderful result of outcrossing Hank's tight DNA. Skipa Star's owner, Sam Wilson, had me fly to Houston and do a shoot. Skipa Star sold for $2.2 million and sired progeny earning over 11,000 AQHA open points.

"Some have neglected the truth that a good rancher is a craftsman of the highest order, a kind of artist."

∽ **Wendell Berry**

The Tahitian Cattle Trade

During my nearly 76 years of life, which have been far better and more than I deserve, I look back at nutty or crazy things, stuff I am proud of, and some other happenings that still seem hard to believe. Bear in mind that fiction must be believable, but truth can just fall where it will—it is what it is. Non-fiction doesn't have to make sense. This non-fiction event falls into the truth category.

A sharp western-looking couple came to the ranch to look at Texas Longhorn cattle and wanted to start a herd. We drove pasture to pasture viewing cattle priced from high to low, talked, and enjoyed a pleasant time. Just when I thought they had selected a group that would be something over $65,000, they dropped the bomb and said they had no money. However, they were in a mood to do some international property trading.

Out of her purse, the lady took a big stack of photos of their Tahiti condo. They would fly to Tahiti to cool their jets during the winters and soak up the pristine blue waters, the music of water rolling against the white beach, and wind softly blowing through the palm trees—all of this enjoyment from their own condo. Their family had been there a dozen or so times and the new had worn off, but it was a great investment and a South Pacific experience like no other place on earth. This condo was right over the water. It slept up to 10 people. The Hilton Hotels & Resorts managed it, maintained it, and used it on a percentage; and the owner received monthly payments. If the owner or owner's guests wanted to spend time in Tahiti, the condo generated no income for that period.

Before the deal was done, Linda was ready to pack bags and head for the other side of the world. Our whole family was excited about this unusual twist for a normal cattle deal. Hilton had provided these people with colorful promotional brochures showing the number of their condo, its location on the edge of the

The Tahiti condo was right on the ocean with the prettiest blue water. The condos were built to reflect a native appearance outside with modern interiors.

ocean, and beautiful photos of the interior. The condos were built out of a kind of Polynesian wood I couldn't recognize, but it was very elegant. Hilton had developed this huge resort and sold shares in each condo with investors all over the world. The setup was very impressive.

The main island of Tahiti was formed from volcanic activity and has high mountains surrounded by awesome coral reefs. It had been a French colony, with the capitol city Papeete. Tahiti is "conveniently located" right on the crossroads of the South Pacific just 3,000 miles south of Hawaii, 6,000 miles from Australia, and 5,000 miles west of Chile. It has a modern airport, and the whole island can entertain the serious relaxer. Tahiti is mostly self-supporting with farms, the sweetest citrus in the world, and panoramic views unexcelled anywhere.

Registered cattle were selling high, but it still took a good trailer load to barter for the Tahiti condo. So we closed the deal, recorded the title, and owned a really southern Southern Division, a world away, thousands of miles to the southwest.

Our family talked about Tahiti, read books, and distributed brochures of our "Southern Division." It was fun to talk about—just another perk for raising Texas Longhorn cattle. Everyone in our family and several employees couldn't wait to get passports. We all continually planned for the day we could get away and go soak up the sun, snorkel, and live it up. But one thing after another happened, and each planned date would be killed by a big cattle sale, a family event, or school sports tournament. Still, we looked forward to the Tahiti experience. It was going to be great.

Winter is the right time to go to the tropics. Because there are 52 weeks in the year, it seemed like a waste to draw the Hilton payments monthly and yet not use the facility, free, ourselves.

In October, we ordered 400 free brochures from Hilton with all the beautiful pictures. We marked each one with an x on our little piece of the Society Islands. A letter went something like this:

> In appreciation of your business with our family this past year, we want to share our personal condo with you as a Christmas gift.
>
> Please check out the enclosed info. Tahiti is the most beautiful group of islands in the world. No other experience will match it. This is our gift to your family. You can fly to Tahiti and stay at no charge for one week. Please select the week of your choice and it is yours, unless that week is already taken. Let us know soon.
>
> A big thank you from our family for your kindness. We hope you enjoy Tahiti.
>
> *Merry Christmas from the Dickinson family*

This gift offer with the invitation letter and a brochure went to our very best buyers. The phones started to ring. Never had we given a Christmas present that was received with the excitement of the Tahiti condo. Our buyers welcomed our gift offer with very warm appreciation cards and calls. We had every intention of coming through with the offer. It seemed like a good idea, and who better to share the pleasant South Pacific than the people who had helped us make the ranch payments, pay salaries, and help make our business successful?

We had enjoyed Red and Charline McCombs' hospitality at the Fiesta Longhorn sales, so they got an invitation. We were happy to offer the condo to them. Charline called and was extremely kind, telling Linda what a generous gift the condo offer was. She and Red would love to accept, but they were really busy. However, they acknowledged the gift with pleasant appreciation. Of course they had their own Lear jet, so it would be simple to just zip right over there for a week.

Calls came in daily during the months of November and December. We were pleased that people appreciated our offer so much, but one by one each declined to spend a week at the Dickinson Cattle Company, "Southern Division." As to cost, we considered at first we might lose several thousand dollars in monthly rental payments, but what is more important than your valuable customers? We would just bite the bullet—it would be worth it to offer this special treatment to our friends.

January came and went and still not a single Tahiti guest. No one wanted to watch the Polynesian dancers do the otea, hear the native drummers beat their shark-skin standing drums, or drink tropical stuff as the evening sun dropped behind Brunei. It was sort of disappointing at the time; but at every Texas Longhorn event we attended, people talked about the "thoughtfulness" of the gift. And it is the "thought" that really counts—right?

We still planned to go to Tahiti, sometime. But it never happened. A couple years went by owning the far-away Southern Division. Then, one of the people who had known of the condo and had received the free offer asked if we would sell our personal Pacific paradise ocean-view property. This was not a buyer we expected to show up at just any given day, but the answer was Yes. We didn't need the condo and hadn't even figured how to enjoy it. So a deal was done, although not for money: we took a trailer load of cows in exchange. So our historic Tahiti condo property came to an end. It wasn't a get-rich deal, but somewhat better than a break-even.

Good Christmas gift ideas are hard to come up with, especially when many people have everything they really need. But every deal brings some useful piece of education. Something always comes of it, good, bad, or otherwise. Actually, the condo was one of the highest-priced gifts we ever tried to give away, and yet it turned out to be the cheapest.

After thinking about the whole deal, Linda and I decided we should just find some very faraway, really nice resort and cut a deal to rent lodging by the week. Get another 400 brochures and send them to all our buyers each year. If no one accepted, we were home free; and if some did, we would gladly pay their week's room rate. Perhaps a log cabin in the tundra of Alaska, a chalet in the Swiss Alps, or a canoe adventure on the high seas of Somalia. All options are still open.

"The pharmacist asked me my birthday again today ... pretty sure she's going to give me a gift."
— **Unknown**

Impressive

I first heard about Impressive when Blair Folck called me and said he had the greatest horse of all time. He flew me to Springfield, OH, and hired me to photograph a stallion. When I got there, I found the "stallion" was just past a weanling, but he had everything Blair had described. I remember being surprised that he would invest a plane fare and photo fees into a weanling colt, but shortly thereafter it all made sense.

Impressive was so spoiled that no one could get him to stand in one place. He was a nearly impossible headache to photograph—an undisciplined, ratty

Impressive, age 2—This photo was taken in the infield at the Ohio State Fair Grounds during the 1971 *All American Quarter Horse Congress*.

infant. We worked and worked with him, but he was on the move all the time. Beautiful, correct, but never in one place for more than a few seconds.

Blair was one of the first "finders" of Impressive, who was never cheap to own. Every buyer struggled to gather his record purchase prices; and the rest of their lives, every seller regretted letting him go. Impressive was said to be "found" by a half-dozen or more keen-eyed horsemen; after everyone else found him, Fennel Brown got to own him, and take him home, and keep him.

Fennel's breeding farm was Brown Quarries, Inc., in Union, MO. His horse-buying funds originally came from construction services

Impressive did not have a weak angle and was easy to photograph. Here, perfect light for a front shot illuminates the veins inside his gaskin. If a horse was too fat, I normally shot from a low angle so it would look taller. If it had white feet, I would darken the foreground chemically so the white feet had defined contrast.

that built indestructible levees all along the Mississippi River. Fennel was a brilliant business man and was paid well for harnessing the mighty, demanding Mississippi River within its banks.

Fennel did not own Impressive—Impressive owned Fennel. The stallion was so valuable, he feared every waking moment that he would die or something bad would happen and his dream horse would go away. He also had a sleepless veterinarian who was prepared to be at Impressive's side for any emergency, day or night. In fact, he built a huge, elaborate suite for Impressive with his own elegant bedroom right next to the stall. A big, thick window between the two rooms let him keep a close eye on Impressive. If the horse made any unusual sounds at night, Fennel would awaken like a clown shot from a circus cannon. While he owned Impressive, he was a happy man; yet he lived on pins and needles 24 hours a day. Some think Fennel's unusual love for this celebrated stallion stressed his heart immensely. Unless he was in the hospital with his repeated heart problems, he stayed close to Impressive.

Posing for a great shot requires a team. Fennel is holding, and his wife, Joann, is working ears. Another fellow would slip in and reset a leg if Impressive moved.

Impressive was a mutation. He was superior to his sire and dam and any of his ancestors, not only in all conformation appearances but also in his own DNA. He was something new and different from his predecessors. Like most mutations, he was a strong breeder; his offspring were characterized by his own exact mind set and consistent championship anatomy. Knowledgeable breeders can recognize the Impressive anatomy stamp for several generations downline.

**Fennel Brown loved Impressive.
Impressive was his life.**

The last time I photographed Impressive was at Fennel's. I had previously photographed him as a 2-year-old when Fennel had just bought him. He was in the Million Dollar Stallion Avenue at the *All American Quarter Horse Congress*. We took him out to the infield of the old race track at the Ohio State Fairgrounds for photos. A group of up to 30 well-known horsemen followed us to the clean, grassy photoshoot area, watching every muscle and every move Impressive made. They knew something wonderful was going to happen with this genetic package. They joked, dreamed, and drooled as they watched. Years later, horsemen who had been there still talked about this photo event with Impressive, the increasingly famous stallion.

Impressive was the 1974 World Champion Aged Halter Stallion. He sired over 2,250 registered foals, of which nearly 30 went on to be World Champions themselves. At one time, his stud fee was $25,000. In other words, if he settled 4 mares a day, he earned $100,000 in stud fees.

Horse people the world over came to Union, MO, to look at Fennel's big red show barn and the famous stallion Impressive—and to look at two famous bedrooms where the horse and the man he "owned" could watch each other 24 hours a day. ▷-D

Impressive at Fennel Brown's, in front of the big red barn

"One hundred years ago everyone owned a horse yet only the rich owned a car. Today everyone owns a car and only the rich own a horse."

BRY Squeeze Chute

Handling cattle is not always easy, or totally safe, depending on personal skills and whether or not the right kind of equipment is used. Early in the Texas Longhorn business, handling was about a horse, a rope, and then throw 'em down for branding. Getting the cow flat on the ground was the method. Later, people designed their own squeeze chutes and panels—even built their own cattle trailers. In order to provide safe ways for cattle handling, every ranch developed its own labor-saving squeeze chutes. Their concern was safety, not only for people but for the cattle as well. Today, ranches continue to design their own systems. Part of a successful ranch plan is being an inventive thinker.

At DCC, we started artificial insemination (AI) in 1973. A wooden gate, squeezed from the side against a board fence, contained cattle for palpation and AI. Doherty 698 had her first insemination in our wood-gate squeeze. The first registered AI cattle in the industry were the result of this wood squeeze. We found that using a head-gate to squeeze their necks was not

This wooden squeeze at DCC was the place of hundreds of AI conceptions during the '70s.

necessary; in fact, cattle handling worked better with a side-squeeze and no head-gate.

In 1978, Powder River designed the Classic Texas Longhorn Chute with the most innovative idea ever—horizontal side drop-out panels. This allowed wide-horned cattle to move freely and without obstruction into the head-squeeze with only two vertical horn-traps right before the head-gate. The industry had never before seen anything except vertical-bar chutes. Powder River's horizontal concept was quickly received, and the company built chutes for leading Texas Longhorn producers all over the nation. Powder River became the main producer of horned-cattle squeeze equipment.

In 1979, DCC designed the first steel side-squeeze chute to safely accommodate the Texas Longhorn's really wide horns. Similar designs that followed were imitations of the DCC design. Many people measured this early design and started manufacturing side-squeeze chutes with slight variations, creating a whole new industry for horned cattle.

In the DCC embryo-transfer barn, we designed a side-squeeze chute for extra-wide-horned cattle. This early model was measured in the early '80s and is still copied by many chute manufacturers even today.

At DCC, over 2,000 embryo calves have been conceived in this horizontal parallel chute that was totally free of vertical parallels—with never a broken leg or horn. Thousands of cattle worked safely in these horizontal parallel side-squeeze chutes over the years. They prove the validity of the idea.

People who handle lots of cattle all understand that they move horizontally. The more horizontal-friendly the design, the easier the flow of cattle. Vertical parallels is a flawed design that can bind a horn or leg and break appendages as cattle move horizontally. Surprisingly, over 98% of all squeeze chutes are designed with vertical parallels that create the most risk for broken needles and hands when cattle jump forward or back. If verticals are used, they should at the very least be spaced much wider than most chutes on the market. But really, there should be no place for vertical parallels in horned or polled cattle equipment.

Most chute manufacturers continue to build heavy, costly squeezes with horn-and-leg-breaking vertical parallel traps. The commercial cattle producers remain comfortable with that dangerous design.

This expensive cattle chute has 34 vertical parallel appendage-traps for legs and horns.

When cows are being palpating or inseminated, the area should always be open below and behind the cow's vulva. Cows will resist when entering a chute; they may lie down or squat in it. They have been known to break instruments and cause serious damage to the palpator's arm.

When attending livestock events, I have always found it educational to drive around the trailer parking lots and look at the equipment. During the Ohio Beef Expo of 2006, Joel and I were looking at the squeeze-chute equipment for sale. I was amazed that one chute was priced at over $13,000; many others were in the $4,000–8,000 range. Considering that the average number of cattle owned by producers in the U.S.A. is 18, it is hard to understand how manufacturers can survive selling these expensive chutes. Not only that, freight alone for delivery on most chutes costs $1,000–2,000.

Over the last 40 years, DCC has built a couple dozen squeeze chutes. Some weigh as much as a Titanic boat anchor. Each concept has been different—longer, shorter, taller, simplified, different spacings, etc.

Our two sons Joel and Kirk put all these ideas together—economical pricing, strength, non-rusting, portable, economical shipping, adjustable, and designed for

The first prototype made by Stein-way Equipment required some tweaking, but it was getting close to right on. Wynn Steiner holding.

widely diverse uses. With our DCC drawings and measurements in hand, seven manufacturers rejected the concept. Then we approached the late Oris Steiner at Stein-way Equipment and found a practical partner to fabricate the new idea—the BRY Chute.

The BRY Horn-Safe Crowding panels were also designed by collaborating with Steiner and his engineering mind. They are a real advancement for appendage safety—very strong, safe, and with great lasting qualities in a permanent corral panel.

We have eliminated all the design flaws described earlier. The BRY is the safest chute for both cattle and people. It is also the most economical of all the steel chutes. Today a complete BRY Chute sells for $1,800, including delivery. More BRY Chutes sell per year than any other squeeze chute manufactured in the U.S.A. The success of the BRY resulted from identifying the problems and inventing an affordable solution. ▷-D

The simple BRY Chute complete. For commercial shipping, it bands down to 9'x5'x7". It fits the U.S.A. herd size for an all-purpose chute. Users state they squeeze many types of stock other than cattle.

"When facing a difficult task, act as though it is impossible to fail. If you're going after Moby Dick, take along tartar sauce."
～ **Life's Little Instruction Book**

Titan Wolf of Kooskia

Watusi cattle are a unique breed developed much like the Texas Longhorn, but probably in a far worse environment—darkest Africa. Watusi are thin-skinned, slick, and well-adapted to the high mountain equatorial regions of the African continent. Some say that Watusi are the largest-horned breed of cattle in the world. They are mostly a dark chocolate-red with huge, very long, upward-turned horns.

Walter Schulz and other family members heard about the huge-horned Watusi cattle which were native to Dar es Salaam, German East Africa (modern Rwanda, Uganda, and Tanzania). Over a 2-year period Schulz shipped 7 bulls and 14 cows out of Africa by boat, landing in Hamburg, Germany, where they were not allowed to off-load. With much effort, a zoo in Leipzig, in the free state of Saxony, allowed entry and placed them in quarantine. From the genetic base of those 21 cattle, Watusi were dispersed as zoo attractions in over a dozen countries.

Pure Watusi arrived in the U.S.A. in the '60s after passing through Florida's Harry S. Truman Quarantine Center. The Schultz family owns the Rare Animal Survival Center (RASC) in Ocala, FL, so the main herd of Watusi in North America was bred there as zoo specimens.

In 1979, Linda and I went to Ocala and bought 15 Watusi from RASC. Nelson Rockefeller and DCC were the first ranchers outside the zoo industry to purchase Watusi. At that time, their estimated population in the U.S.A. was only 48, and DCC owned nearly a third of that number. Since then, the breed has expanded and is in private herds and zoos all over many countries.

Due to intense inbreeding among the first few shipped out of Africa, all Watusi outside of Africa continue to be even more inbred. An up-breeding, out-crossing program has actually evolved to 15/16 and 63/64 Watusi, which has freshened the pedigrees with slightly different DNA. By starting with half-breed stock, then breeding them to pure again for a 3/4 Watusi, the percentages were built—ever continuing higher.

DCC collected the first Watusi semen and performed the first successful embryo tranfers. Although the Watusi herd has remained mostly from 10–20 head, summer tourists especially enjoy seeing this boldly striking family of exotic cattle.

DCC has searched for superior Watusi genetics using numerous bulls and AI. To help create percentage up-breeding, we have exported semen from Akeem, Buffalo Bill, and Titan Wolf to New Zealand and Australia. Some very fine specimens are being developed.

After breeding Watusi for nearly 40 years, we have found that people especially like the very wide, flat-lateral spreads and the wild-spotted African cattle. That type sells best. However, finding breeding stock of this novel combination is a very hard go.

A few really good Watusi are scattered all over. By 2015, I had been looking for a distinct, unrelated bull for a long time. Duane Gilbert of Castle Dale, UT, pointed me to Titan Wolf for sale in Kooskia, ID. He was a direct son of Olympian, a lateral-horned bull raised by the Schultz Catskill Game Farm in New York.

Mel Hedberg, who owned Titan Wolf, was a retired Idaho Fish and Game officer living in the colorful Clearwater Valley. He had purchased his foundation Watusi herd from Mad Dog Ranch, owned by famous rock musician Joe Cocker. When Mel sent photos of Titan and a number of his progeny, I could see this was the exact bull I wanted. We made a deal.

Kooskia, ID, is in the high mountains just below the Canadian line, exactly 2,197 miles from DCC in Barnesville, OH. Mel and Titan Wolf lived up a long mountain road. I never seem to locate a special bull anywhere near our Ohio ranch.

Titan was entering his 13th year, and his age was a concern. Mel agreed to take a partial payment, and when (if) Titan proved to be a successful breeder, the balance of the purchase price would be due. I had never met Mel, but I thought that offer was very fair.

I called dozens of transporters with little success locating anyone to pick up Titan. Finally, I found a rodeo transporter who was moving a load of bucking bulls from British Columbia to Texas and hired him to make the trip through Kooskia to pick up this big-horned African bull. Titan rode to Texas with a load of bucking bulls. Shop talk could have been wild if you were into traveling-bull chat.

Titan went from Idaho's high chilly mountains down to Champion Genetics in Texas, where he was greeted by 105°. The climate change knocked a chunk out of him. He went sterile. Cattle don't just take off a coat on a hot day—they take time to adjust. For three months, Champion collected Titan weekly with no success. He adapted slowly to the hotter environment, and quality semen was finally processed in early fall. After months of testing, his semen was offically qualified and has been exported to multiple countries.

The search for superior genetics is not easy. With minor breeds like Watusi, the numbers to select from are small and very difficult if one is picky about quality.

Titan is getting older and still breeding; his beautifully spotted Watusi calves are populating pastures all the way from Ohio's Appalachian hillsides to Australia and New Zealand. He is being used on daughters of Mr. Immambo, Buffalo Bill, and Pretoria. The excitement of raising the very best is always rewarding, no matter the risk.

Titan Wolf, age 14, at home at Dickinson Cattle Company. He has calves on 3 continents with records of ancestors in Africa going back over 6,000 years.

And, yes, Mel did get full payment for the balance, somewhere up there in Clearwater Valley among Kooskia's rugged mountains.   ᗪ-ᗪ

When the first crop of Titan Wolf calves arrived, nearly every one had his "Swede" spot pattern: dark on the top and white on the bottom half, with a spotty little design in the middle.

"Excellence is the slow, gradual result of always striving to do better."

∼ **Pat Riley**

Drag Iron

In 2002, Tom Smith of Michigan bought Dickinson Cattle Company semen from the grand old bull Shadowizm. He bred his good cow Jamoka; and in 2003 a long, tall, dark-colored bull was born, later to slick off to be the darkest brindle ever. Tom named him WS Jamakizm. His neighbor Dick Lowe liked the looks of this young bull and bought him.

In 2005, while my wife, Linda, and I were taking advantage of Dick and Peg Lowe's hospitality, Dick took us to a pasture north of his house. He showed us a tall, dark, young bull who was doing incidental surveillance of cows in a thick-wooded pasture. This was our first view of Jamakizm, and, wow, was he a nice bull! I wanted to buy semen, so I encouraged Dick to collect this bull—and soon.

Several months later, Dick called. He had semen available for $100 per straw. How much did I want? At that time, $100 was top price for Texas Longhorn

Drag Iron is a striking bull who catches the eye of everyone who looks at him. His size, color, gentle disposition, and thickness demand approval.

semen. I thought his price was high, but he stood firm. I knew the cost of collecting semen and also what Dick had paid for the bull, so I offered him $5,000 for 100 straws. I pointed out that my $5,000 would pay for all his collections as well as what he had paid for the bull. He was still firm—for a while. However, eventually he agreed to take the offer.

In 2006, DCC used Jamakizm's semen through AI on a variety of cows. He was an unproven bull, so every mating was a calculated guess. The following year, 28 Jamakizm calves were born at DCC. The first was Juma, followed by Drag Iron a few days later. Both were wild crazy brindles, built tall and long like their sire, yet thick like their dams. Shortly thereafter came Jama Dandy, Play The Jam, Jamster, Jamaju, Toss The Jam, Jack Pot, Hooray, Jam Packed, and more. At the very first glance, it was obvious that Jamakizm progeny had a winning look; he soon proved to be one of the strongest-value sires in history.

Like an Indian rain dance, timing is everything. Drag Iron and his whole dark-brindle family came into an industry that was flooded with white bulls. People were looking for genetics to add frame and thickness and to darken up their herds. Drag Iron did just that.

Pretty Lady by Drag Iron sports 94" horn tip-to-tip and is only age 8.

Silent Iron by Drag Iron was the 2017 Horn Showcase Champion with 91" tip-to-tip at age 4.

Few would disagree that Drag Iron is the prettiest colored brindle ever. His impressive size pushes the scales down to 2,260 lbs. His mature 87" tip-to-tip spread puts him into the upper-college class for horn. His daughters and sons win futurities and shows. His daughters have topped several sales, and he is still a young bull.

After natural breeding, artificial insemination (AI), and embryo transfers, Drag Iron's genetics fill the DCC pastures. In the fall of 2016, we sold Drag Iron to Glendenning Farms of Celina, TX, but DCC continues to sell his semen. Drag Iron has progeny in New Zealand, Australia, Canada, and the United States.

One great bull like Drag Iron has the ability to produce hundreds of calves to bless the coffers of many people who utilize his DNA. He is a result of inspired selection and careful matings. He is a planned design of 12–18 generations since the first registered Texas Longhorn cattle. Breed improvement comes slowly, but great value is the result—for those with years of patience. ▷-D

> *"The truth is like a lion. You don't have to defend it. Let it loose. It will defend itself."*
>
> ～ **St. Augustine**

Clear Point—The Linebred Outcross

Clear Point is not a mutation. A mutation is "a sudden departure from the parent type in one or more heritable characteristics, caused by a change in a gene or chromosome." Although Clear Point has extreme, individual characteristics, he is not a "sudden departure," nor is he much different from his sire. His originality is due to more than a hundred functional individuals in his ancestry that through the generations gradually improved over the foundation of their origin.

In 1967, when the first registered Texas Longhorns were weighed at DCC, and later, as other cattle walked onto the ranch scales, we were amazed that many adult cows out of south Texas were only 550–700 lbs. True, they had the longevity,

This is Clear Point, age 5, weighing in at 2,245 lbs., many times Horn Showcase Champion, with over-91" horns tip-to-tip. He is the pinnacle of DCC's linebred outcross genetic plan.

calving ease, disease resistance, and browse ability, but they were small. Commercial cattlemen are concerned about their ability to gain weight. The unprecedented Texas Longhorn attributes are unquestioned. The only thing left to achieve is holding together this basket of virtues and adding good gainability to the genetic package.

The year 2017 was the 50th that the Dickinson family bred, raised, and performance-tested Texas Longhorn cattle. Small increments of improvement have come painfully slowly. Little by little, better-gaining specimens developed that still held onto the old proven Texas Longhorn virtues.

A herd sire is like a toolbox. Each job requires a different tool. A saw is wonderful for cutting boards but a wreck if used to drive a nail. Hammers can't make a straight cut. Bulls are like tools, each providing his DNA to sire certain characteristics. Some bulls offer many desirable characteristics while others have a one-tool virtue. People ask who my favorite bull is, so here is my toolbox answer: it depends on the genetic job that is needed.

Clear Point is a result of over 60% artificial-insemination and embryo-transfer-conceived ancestors. He has 126 predecessors with DNA or bloodtype data on

Clear Win is the sire of Clear Point. He is a leading sire of over-90" sons and a many-times Horn Showcase Champion.

file at the TLBAA and ITLA association offices. His ancestry is amazingly easy to research.

Clear Point is the result of the DCC linebred outcross banding practiced for over 40 years. His pedigree shows the linebreeding of breed-leading genetics, outcrossed with other linebred select individuals. He traces 71 times linebred to the Phillips bull Texas Ranger JP. He is 24 times to Don Quixote, 10 to Measles, 8 to Overwhelmer, 7 to "King," and 6 times to Senator. He is not a mutation, just a nice genetic bounce above the data of his carefully selected ancestors. Texas Ranger JP, Clear Point's multiplied ancestral line, is documented by TLBAA as appearing in pedigrees 58,626 times—more than any other individual of record.

Little pieces of breed improvement add up, inch by inch, pound by pound, data on top of data. Clear Point's breeding reads like a book and writes the story of DCC efforts. ▷-D

Clear Point has unequaled frame and muscle structure.

"Everybody pities the weak; jealousy you have to earn."
∼ **Governor Arnold Schwarzenegger**

Power Game—The BueLingo

The BueLingo breed of cattle was developed in Ransom County, ND, on the Bueling Ranch, by Russell Bueling and Dr. R.B. Danielson of the Animal Science Department of North Dakota State University, Fargo.

These two men founded the breed by following a planned program to develop cattle that suit today's needs for low birth-weights, fast growth, extra milk, and quality carcasses. As the years passed, observation of the breed showed that the BueLingo possessed traits exceeding those of many other breeds. Their success is evident as the breed is now well-accepted all over North American and is being perpetuated in 28 states and 3 Canadian provinces.

Their distinct "wow" appeal, such as their novel color, combined with great beef attributes, gave DCC reason to make them the third breed of the ranch's registered breeds.

Buying the best stock possible was difficult in the late '80s; producers were keeping their quality stock to further develop their own herds. DCC was forced to buy pure Dutch Belted dairy cows to get the strong belts, then band this with semen from registered BueLingo bulls and other leading performance cattle such as Salers, Limousine, Charolais, and Angus. With 100% performance testing from the beginning, superior-gaining BueLingo were developed with a cautious eye on calving ease. We fine-tuned the breeding herd by tracking data on all fed steers. And we weigh every calf at birth, and don't keep objectionably large birth-weight cattle for breeding.

In 2009, a major genetic jump came with the birth of Power Game. He was a low-birth-weight, perfectly belted, black bull from a two-year-old heifer. At 205 days, he was adjusted to 758 lbs. with BIF (Beef Improvement Federation) for a 2-year-old dam. With this great data, comparable to the best of other fast-gaining breeds, we used Power Game heavily to further develop the DCC BueLingo genetics, collected his semen, and made it available to the public.

Power Game gave the BueLingo herd a whole new authority with his great numbers, totally above the previous leaders. His 758-lb., 205-day weight has moved the profit forward.

During the summer tours, BueLingo make a real show for guests. The cattle hear the bus horn and come at a run for a few hand-fed cow-candy treats.

Power Game—The BueLingo

Due to the love of livestock, many cattle are raised not just as a camouflage piece of meat. Often special producers' druthers are for critters flamboyant in the pasture—a conversation piece, something different, and a stand-out breed to be especially proud to raise.

Today, semen and breeding stock from our BueLingos are sold nationwide. DCC's own herd runs just over 120 females, has striking belts, low birth-weights, and great early gain. The brood cow herd is 37% sired by Power Game, and many of the younger generation are grandsons or daughters. The DCC BueLingo efforts are nearing 30 years, proving that 100% performance testing is the secret to exponential success. ▷-D

"God makes a way out of nowhere."
— Rev. Norvel Goff, Sr.

Field of Pearls, the Family

Johnnie Hoffman of Metairie, LA, was a brilliant equipment specialist in charge of many of Noble Oil's off-shore drilling efforts. People from Noble said Johnnie could start up any kind of huge oil-drilling equipment and make it function at its best. He knew how to drill the depths of the Continental Shelf off the coast of New Orleans in the Gulf of Mexico. In fact, today one of the largest Gulf drilling platforms is named "Noble Johnnie Hoffman."

Johnnie was a profitable manager for Noble; when he tried to retire several times, Noble begged him to stay by offering him sizable raises. While his body was sweating under a hard hat, his mind was on race horses and Texas Longhorns. As his appreciation for Longhorns increased, he liquidated his horses.

He developed his Longhorn ranch near Lake Pontchartrain, LA. The ranch was pretty with coastal grasses, turnips for winter livestock grazing, and huge trees with hanging Spanish moss.

Just as Johnnie had a serious eye for good running horses, he was every bit as serious about Texas Longhorns. In the late '70s, he started buying very good cattle from DCC—over 20 of our high-dollar embryo heifers, a number of pregnant recips, and 2 bulls, Texas Freckles and Emperor. From a DCC embryo recip, he raised Dixie Hunter, probably the best Classic son.

Johnnie worked to make the Louisiana Purchase sale at Covington, LA, a high-quality annual sale. He was part of an effort to sell about 100 cattle. The sale was very well done. I found a cow there named Impression, with the frame, correctness, substance, a smoky speckled-brown color, and perhaps a larger size than the mostly white cows in the sale. She had baggage. Impression's udder was lop-sided, and her value was further reduced by a broken horn tip. I bought her for $2,050. She was a cosmetic cull, yet a genetic diamond.

In 1997, Impression had a Zigfield bull calf that never enjoyed enough milk, as his aging dam was a low-milk producer. Many cows that give a lot of milk tend to

Field of Pearls

have quarters go blank or reduce milk production over the years. Her calf was a dogie, never slicked off, and had a dead-hair coat for over a year. Although he had one of the smallest weaning weights of the bulls, he had a special look and caught my eye for some unknown reason.

Impression's son was named Fielder. As a two-year-old, he was given test cows; his very first calf was a strapping big heifer named Field of Pearls. Over the years, Fielder became well appreciated as a maternal sire of great producing females. They had the ability to produce qualities noticeably superior to themselves. Bowl of Roses, for example, was the dam of Clear Win, a Horn Showcase Champion, out of a Fielder daughter.

Field of Pearls was beautiful. As a yearling, she was ITLA Junior Champion, Non-Halter Female. Later, in 2006, she was judged All-Age Grand Champion ITLA Non-Halter Female, going head-on against heavy competition. Her years of planned matings developed a linebred outcross blend pedigree of 16 times to Texas Ranger for beef, horn, and size, then 4 times to Measles for trim correct type. She became the best of all the predictable good stuff.

We bought Over Head semen from Burton/Stockton Ranch. Field of Pearls was first in line to be bred and produced the prettiest black-speckled bull calf. He was born on March 3, 2003, and we named him Over Kill. After years of service, he was injured and passed away before his full time; yet in his short life, he was productive enough to sire several hundred calves, mostly black-spotted. He was the first black-spotted bull to measure over 80" tip-to-tip. Over Kill was known for siring very straight backs, trim underlines, and a low, twisted spread he personally carried with a 2,100-lb. body. His progeny were black-based or mostly dark colors.

When we acquired Temptations The Ace's much-sought-after semen, Field of Pearls was the first cow to receive it. Her highly coveted calf Tempter was born April 17, 2005. Both Tempter and Over Kill, almost totally different bulls, were bred at DCC. In addition to their AI breedings, they had a total of 18 years of natural service at DCC—Tempter, used 7 years and Over Kill, 11 years. These two bulls created a pleasant annual cash flow for the ranch with their progeny; they made Texas Longhorn ranching a pleasure. Today, hundreds of prominent cattle carry the Field of Pearls blood through these two great sons.

Over Kill as an unweaned calf

In 2011, Field of Pearls produced a huge heifer, sired by Drag Iron, which we registered as Dragon Pearl. As amazing as her dam's produce record was, Dragon Pearl's appears to be even greater. Her first three calves are Reckon So, Time Line, and Cut'N Dried. All 3 are projected to easily exceed 80" tip-to-tip while still young. Dragon Pearl weighs 1,545 lbs. She is extreme in both her horn development and body size.

Field of Pearls has 20 calves registered in ITLA, most still retained by DCC, and over a hundred granddaughters in the herd. She has replaced herself with her younger family. In 2012, Justin Rombeck negotiated purchase arrangements, and we sold this famous cow to the Todd McKnight/Alex Dees Partnership.

Under this new management, Field of Pearls has already had 19 calves registered in TLBAA, all by embryo transfer. The most famous, Tuff Stuff, has surpassed the 80" benchmark and is projected to be the first Tuff son to go over 90". Field of Pearls has the ability to be bred to a wide assortment of unrelated bulls and produce breed-leading males and females. Her unusual diversity is enormously valuable.

Every registry has leading specimens that hold exciting promise, yet never produce a single offspring that matches or exceeds their own quality. Field of

Over Kill at 18 months

Pearls consistently produces over her head. To identify these individuals at the earliest age and capitalize on their strong DNA is to harvest untold assets. The process often takes 6–10 carefully evaluated calves to determine if a cow can out-produce herself. This is a case in point for the serious breeder who pursues a planned program with systematic data evaluation. It is one thing to raise a Field of Pearls, one thing to mate her correctly, and yet another to financially plan her genetics for the next forward jump.

As Justin Rombeck says, "Her story is far from over—she will live on through her offspring and the impact they will continue to make for generations to come."

Field of Pearls and her family were bred and raised by Joel Dickinson as part of his personal herd. ▷-D

Over Kill—mature, beautiful, and striking

"Oregano and tomatoes make it Italian. Wine and tarragon make it French. Sour cream and vodka make it Russian. Lemon and cinnamon make it Greek. Soy sauce and duck make it Chinese—but grass-fed Texas Longhorn beef makes it healthy."
∼ **Darol Dickinson**

Dickinson Cattle Company — One Month

Linda and I were married April 21, 1963. When Linda let me take her out for the first time, I was age 19 and she was 17. I had been praying that the Lord would help me make a right decision on the correct wife, so I decided to propose on the first date. She was better than I deserved. How could I lose? But before I returned her to the safety of her parents' home, I thought, "Wow, I don't want to mess up and scare her!" so I gave it some time. The following Saturday night, I decided it was time—the last thing I wanted her to think was that I was indecisive and couldn't make an easy, no-brainer decision. So I proposed on the second date.

Two years after the wedding, we began having kids on a regular basis. More than 50 years later, we enjoy having grandkids all over the place.

You have followed this drama from locating $1,000,000 cattle in the swamps of Texas, being ripped-off by friend and foe, hunting the mountains of Idaho for a novel bull, walking the racetrack shedrows hunting a perfect Thoroughbred, getting run out of restaurants, and rubbing shoulders with the likes of Carl Miles,

The main business for DCC is producing and selling quality registered cattle. It is a great joy to see a load depart the ranch for a new buyer to start a herd.

This is our family, Christmas of 2017.

Sheriff Buck Echols, Elmo Favor, Charlie Dees, Walter Spencer, Hank Wiescamp, Walter Merrick, J.G. "Jack" Phillips, Frank Doherty, Pauline Russell, and Frank Dickinson, my dad.

As you complete this reading experience, this last chapter is my wrap-up. Here is what typically happens at DCC, in a month. This is the status of a little boy's dream and a totally innocent little ranch girl's marrying an optimistic fellow with no money.

Ecclesiastes 9:11: "I returned, and saw under the sun, that the race is not to the swift, nor the battle to the strong, neither yet bread to the wise, nor yet riches to men of understanding, nor yet favour to men of skill; but time and chance happeneth to them all."

Realizing that Solomon's wisdom in the 10th century B.C. was totally true, I knew "time" and "chance" were my best resources to work my dream for ranching. I needed to put in long, hard hours, try a lot of things, and take a lot of chances to achieve my goals. You have read about my decisions and efforts. You can decide for yourselves if I acted right or wrong.

Ranching as a business is distressed. As taxes and government regulations exponentially squeeze, the huge landholdings are dwindling. Over 1,000 ranches in the U.S.A. are going under per month. However, the lure of grass, livestock, clean air, and pure water refreshingly hangs in the air. New young ranchers are ready to try their hands.

Young men and women coming up in ranching are being confused by unhelpful college training. Very few college professors own a single cow—but ranching is an industry where you have to own it to understand it. Colleges teach science and management but fail to teach livestock retail marketing. Good management is fine, but great marketing will cover a multitude of management errors.

Dickinson Cattle Company is a family thing. Joel, our youngest son, oversees everything that eats grass, the veterinarian visits, the embryologist, the land, vehicles, cattle sales/purchases, employees, and a multitude of unknowns that just pop up without notice. Kirk, our oldest son, is in charge of everything that causes electronic problems, computers, surveillance cameras, internet marketing, online store sales, drones, publishing the *Texas Longhorn Celebrity Calendar*, tour-bus

sound systems, book design and layouts, tax harassment, government forms, and almost everything that is supposed to be user-friendly in the modern age.

Longhorns Head To Tail Store (LHTT) is open 10:00 AM to 5:00 PM except Sundays. The store was designed to market all natural, grass-fed, and grain-fed beef grown here at the ranch. We sell several thousand pounds of lean grind and freezer-beef halves each month—all vacu-wrapped and government inspected. Longhorns Head To Tail Store began with only beef sales, but has expanded and now has a wide range of inventory "you can't find at Walmart." Every day, the store features hair-on speckled steer hides, taxidermy head mounts, bulk foods, choice thick steaks, great books, ranch logo clothing, polished western skulls, #6 bear traps, semen storage tanks, Longhorn kid toys, the BRY Chute, cattle scales, and 75 different horn products.

Beyond polished skulls, Joel and his family have developed many beautiful uses of horn. From small, very colorful, polished horn pieces, they make handsome bowls and elegant jewelry such as earrings, necklaces, hair-horn décor clips, bangles, and bracelets.

Tour buses arrive near the ranch pavilion and take ranch buses for a bumpy pasture tour. Several thousand guests per year enjoy learning about ranching, history, and all kinds of livestock.

Another unique feature of LHTT is its Longhorn Tour program. From May through October, several thousand people ride ranch buses for the 75-minute narrated tour. This ranch tour takes guests bouncing over miles of beautiful, rolling, grass-covered Appalachian hills where they will see hundreds of cattle from the championship Texas Longhorns, African Watusi, and Dutch BueLingo breeds. Guests hand-feed cattle from the buses. The tour gals answer frequently asked questions about how many acres of land make up the ranch and how many cattle make up the herds. They also share local history, including that of famous and colorful past residents such as Hanging Judge Isaac Parker and William Boyd—Hopalong Cassidy. Guests learn about cattle breed histories, the virtues of high Omega-3 beef, the billions earned locally by coal miners of old, and current efforts to drill horizontally for oil and gas in this region.

An option for some Longhorn Tours includes a catered meal served in the ranch pavilion. The summer tours require extra cast members to help the guests. Normally our staff includes 10 employees. In the fall, our annual Customer Appreciation Day draws invited guests who have purchased cattle. They enjoy a day of demonstrations, educational seminars, and great food.

DCC ships semen mostly in the U.S.A., but also worldwide. These sales are handled daily by my hard-working, silver-haired wife. In 1988, DCC set a goal to establish herds of Texas Longhorn cattle in as many nations as possible. To date, we have exported live cattle, embryos, or semen to 26 countries including Australia, Bolivia, Brazil, Canada, Costa Rica, the Czech Republic,

Embryos were shipped frozen to Israel and implanted into native cows. Next year, Texas Longhorn calves originating from DCC will be born in 26 countries.

Joel is measuring Shadow Jubilee, one of the first over-90" T2T cows. He does the AI, measures and weighs the cattle, and manages everything outside.

Denmark, Germany, Israel, Mexico, New Zealand, Nigeria, Poland, Sweden, Switzerland, Thailand, and The Netherlands. We get excited when importers send photos of their newborn calves from DCC frozen-embryo science.

In the spring, Joel inventories every pasture to locate new calves. Each calf receives a selenium shot and is weighed, some are band-castrated, all are ear-tagged with the sire's and dam's names printed on each tag (no number tags—this isn't government work). Each calf is also photographed. This info is emailed direct to the office where the silver-haired lady records all data. That same hour, a computer file is born for each calf and continues to record its every life change, movement, breeding, weight, and at last a sale price.

Joel's wife, Misty, works for a local doctor. Sometimes her nurse training is helpful at the ranch. When someone gets an injury, her diagnosis usually boils down to "You can walk it off." The cowboys do get banged around some, but they hardly ever go to the hospital.

Roger Moore began working with DCC soon after the ranch settled in Ohio. He is a mechanic, cattleman, farmer, everything, and drives a wicked Honda

Kirk and Linda "Teu" Dickinson with nephew Bry and their oldest granddaughter, Jasmine

all over the ranch hundreds of miles a month, filling mineral troughs and feeding cattle. If a steer gets out on the highway at 2:00 AM, one call to Roger and within minutes he will be on the job resolving the problem. We appreciate Roger very much for his years of loyal service to DCC.

Really messing things up, Kirk married a Linda. Having two "Linda Dickinsons"—my wife and my daughter-in-law—working on the ranch really confuses the phone system. Therefore, we have affectionately renamed Kirk's Linda, as "Teu" Dickinson. If everyone gets it right, she is the number two of the Lindas. Teu moves the meat and delivers frozen beef to clients. She is the ace tour narrator and really plays with the minds of guests. Her bus passengers return from the pastures just short of rioting, stagger off the bus, and wilt into the ranch store. She makes every tour an entertaining experience.

Our daughter, Dela, represents the ranch at cattle events in Texas and Colorado. She comes from Colorado to Ohio several times a year to help with special events such as the fall field day. Her husband, Steve, is a master electrician and has wired most of the buildings at DCC. He is the creator of our beautiful automatic nightlight system. When a coyote waltzes by the pheasant aviary in pitch darkness, the flood lights come on.

Chad, our middle son, and his kids attend Longhorn events in Texas promoting DCC and meeting a lot of people. He designed and built all the buildings at DCC. Since creating his first seedy tree house, he has loved construction. We will

The last Saturday of September, DCC stages a Customer Appreciation Day for all previous cattle buyers and special guests. This is the prime-rib-serving, family, demonstration, seminar team 2017.

always need more buildings. He is a master framer doing custom homes in the society area of Austin, TX.

During the summer, 2–6 grandkids work on the ranch. They all work in their areas of expertise. Brice, Trista, Austin, Bry, Shane, and Kara work, sort, and feed cattle, drive herds, and halter-train young bulls. In addition to working with the cattle, they build fence, brush-hog, run for parts, keep records, paint fences, and eat. Eating is where my silver-haired wife excels as she prepares great rib-sticking lunches every high noon—lots of beef. Trista helps Linda with the food preparation.

Rose Arend is from the Flying Arends, a professional acrobatic water team. She is in charge of the LHTT store, tracking inventory and displays, shipping mail orders, and booking and narrating ranch tours. She also oversees roof repairs for our buildings and oil changes on the buses. Rose promotes LHTT at Chamber of Commerce meetings, tour promo expos, etc. Happily, she breaks fewer bones than in her previous job.

In the early spring as new calves are born, we decide matings for the following breeding season. Since a wrong decision costs a year of the cow's production, we evaluate every calf to determine if the correct bull was matched to the right cow. About 60% of the cows are placed with a different bull each year as the search continues for perfect matings. Sixteen bulls are bred by single-sire, natural service. This planning takes a lot of time for Joel and me. We compare notes. We recall current and past combinations that worked well or were costly failures. Before we make final decisions, we check matings and review them against Linda's computer data. This time-consuming study lasts from March to June 20, then bulls go with the cow herds starting in mid-June.

Each year, we select certain outside bulls for frozen semen testing, mostly for bringing in fresh, unrelated blood. This testing is hard, because very few breeders in the industry document weights and measurements with the same detail as DCC. When buying semen, we find that documentation on new bulls is often vague. Many times, we test new bulls from outside bloodlines in hopes they will give a breed advancement beyond DCC genetics. Going outside DCC genetics almost assures a loss of size; but regardless of quality, outside bulls will break the pattern of linebreeding proven to work on our cattle. We select about 120 cows for artificial insemination—over half will be AI bred to certain proven matings of DCC bulls.

TLBAA awarded the Dickinson family the Lifetime Achievement Award for breeding and promoting Texas Longhorns.

Our AI season is in June and July. Doug Burris and Kara Dickinson have attended AI school and can do the job. With some $500 semen, Joel still wants to be in charge and AI those personally. All three watch cow cycles. With a professional AI program, there is no room for error.

Kirk developed the first comprehensive computer program for documenting cattle data to show pedigrees, weights, and horn measurements with total performance info. When he began building the program, no systems included all the data that DCC needed to track. Today, the ranch site **www.texaslonghorn.com** can trace pedigrees back more than 14 generations with multiple comparative photos and data. Buyers purchase more cattle from the DCC online site than from most registered programs of any breed. Online, DCC also offers over 400 educational articles to help all producers.

The good team of family members manages all parts of the ranch. Looking down the road, a half-dozen grandkids are in the wings as possible successors 20 years from now. Running a ranch has to be in your blood. It's not a govern-

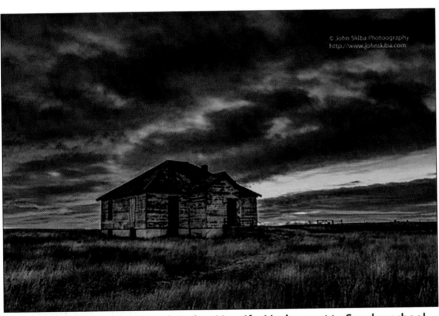

This is Edison Leader Church today. My wife, Linda, went to Sunday school and learned moral values here. She ran around this building at Bible school, laughed, played on the swings, and skinned her knees. It is slowly rotting on the vast prairie lands near our former eastern Colorado ranch. The church witnessed tears, joy, weddings, and funerals. The grave yard is behind the building. Photo Credit: John Skiba Photography

ment job with holidays, 8-hour days, and time off. When cattle need feed and care, someone must be watching. Ranching is a big dedication.

Friends tell us we should go on a vacation. Go to Florida in the winter, or to south Texas. We think about it. We thought about going to Tahiti. After some thought, we realize that for the past 50 years we have worked to build a ranch, environment, and location where we want to be. We like living in the middle of the grassy, rolling hills of Appalachia. We like the climate, the lakes, trees, cattle all around, neighbors, and most of all, family. It is a joy for family to be near and see grandkids fly by on a Honda scooter or show up at the ranch office hunting food. Looking out the office window, I enjoy viewing a herd of cows on the horizon of our "Mexico" pasture. The herd sires are in pipe corrals right below the office deck. All things considered, we can't think of any other place we would rather spend our time than right here.

So here we are, enjoying days that are full of variety. Daily we take calls from customers ordering products. Ranchers call about buying cattle. Folks show up to receive cattle they have bought. We provide interstate health papers and photograph ranchers and their loads to record their special buying day. They may be a young couple starting their herd—that is a big deal. (We remember when we did that ourselves.) Joel calls and reports that a new calf from a special cow really looks good; it is standing, nursing, and has been tagged. Or maybe he tells us that one of the valuable bulls is limping. He will be driven to the corrals and dealt with. Families come to the store and buy freezer-beef halves. I look out the office window as the ranch hands drive a pasture of cattle down Muskrat Road. A tour bus is slowly moving in the middle of the herd. I can hear the tour guide's narration as the herd treks by.

I don't know how to make it better. As Charles Goodnight said, "When the ranch is at peace, no other life is more perfect." ⊳-ᴅ

"And also that every man should eat and drink, and enjoy the good of all his labour, it is the gift of God."

~ **Ecclesiastes 3:13**

A Word of Praise

I am not a good writer. Good authors can write convincingly about things they have never seen or lived. I am not that professional. A fiction writer must brilliantly create the plot, create the twists and misconceptions, then tie a bow on the adventure and make it, in the end, sound believable.

I just write about things I have done, seen, or personally viewed. Non-fiction doesn't have to have a bow. It just happens and there it is, stark naked. It don't have to make any sense at all.

To keep me from appearing to be a third-grade drop-out, a big Thank You to spellcheck, Google, Wikipedia, my computer son Kirk, and proofreaders Mark Cooper (Antrim, OH), Professor David A. Richardson (Cleveland, OH), and **Western Horseman** publisher Randy Witte (Peyton, CO). If there is any class or professionalism—it's all their fault. DD